Mastering
ROBOTICS
AND AUTOMATION
Concepts, Techniques, and Applications

Nikhilesh Mishra,
Author

Website
https://www.nikhileshmishra.com

Copyright Information

Copyright © 2023 Nikhilesh Mishra

Dedication

This book is lovingly dedicated to the cherished memory of my father, **Late Krishna Gopal Mishra**, and my mother**, Mrs. Vijay Kanti Mishra.** Their unwavering support, guidance, and love continue to inspire me.

Table of Contents

Author's Preface

Welcome to the captivating world of the knowledge we are about to explore! Within these pages, we invite you to embark on a journey that delves into the frontiers of information and understanding.

Charting the Path to Knowledge

Dive deep into the subjects we are about to explore as we unravel the intricate threads of innovation, creativity, and problem-solving. Whether you're a curious enthusiast, a seasoned professional, or an eager learner, this book serves as your gateway to gaining a deeper understanding.

Your Guiding Light

From the foundational principles of our chosen field to the advanced frontiers of its applications, we've meticulously crafted this book to be your trusted companion. Each chapter is an expedition, guided by expertise and filled with practical insights to empower you on your quest for knowledge.

What Awaits You

- **Illuminate the Origins:** Embark on a journey through the historical evolution of our chosen field, discovering key milestones that have paved the way for breakthroughs.

- **Demystify Complex Concepts:** Grasp the fundamental principles, navigate intricate concepts, and explore practical applications.

- **Mastery of the Craft:** Equip yourself with the skills and knowledge needed to excel in our chosen domain.

Your Journey Begins Here

As we embark on this enlightening journey together, remember that mastery is not just about knowledge but also the wisdom to apply it. Let each chapter be a stepping stone towards unlocking your potential, and let this book be your guide to becoming a true connoisseur of our chosen field.

So, turn the page, delve into the chapters, and immerse yourself in the world of knowledge. Let curiosity be your compass, and let the pursuit of understanding be your guide.

Begin your expedition now. Your quest for mastery awaits!

Sincerely,

Nikhilesh Mishra,

Author

CHAPTER 1

Introduction to Robotics and Automation

Robotics and automation have transformed the way we interact with the world, revolutionizing industries, enhancing efficiency, and opening up new frontiers of possibility. In this chapter, we embark on a journey into the heart of these dynamic fields, where we will unravel the core principles, historical evolution, and fundamental concepts that underpin robotics and automation.

From the most basic definitions to the intricate web of technologies and applications that make up this domain, we will delve into the world of sensors, actuators, control systems, and the myriad ways they enable machines to mimic and even exceed human capabilities. We will explore the rich tapestry of robotic applications across industries, from manufacturing floors to outer space, and gain a deep appreciation for the benefits and challenges these technologies present.

Whether you are a seasoned professional seeking to refine your expertise or a curious newcomer eager to grasp the fundamentals, this chapter will lay the groundwork for a comprehensive understanding of robotics and automation. Join us as we embark on this journey, where the convergence of science, engineering, and innovation propels us into a future where machines and humans collaborate seamlessly, bringing about remarkable advances in our world.

A. Definition and Scope of Robotics

At the heart of any exploration into robotics lies a fundamental understanding of what robotics entails and its scope in today's world. Robotics represents the synergy between mechanical engineering, electronics, computer science, and artificial intelligence, culminating in the creation of machines capable of performing tasks autonomously or semi-autonomously, often with a focus on interacting with the physical environment.

Defining Robotics: At its core, robotics is the science and engineering discipline that deals with the design, construction, operation, and application of robots. A robot, in the context of robotics, is a machine designed to carry out tasks, manipulate objects, or perform functions with a certain degree of autonomy. This autonomy can range from simple, pre-programmed actions to advanced decision-making based on sensory inputs and artificial intelligence algorithms.

Scope of Robotics: The scope of robotics has expanded dramatically over the years, encompassing a wide array of applications and domains:

1. **Industrial Automation:** Perhaps the most well-known and established domain, industrial robotics involves the use of robots in manufacturing and production processes. These robots perform tasks such as assembly, welding, painting, and material handling, often with precision and speed far surpassing human capabilities.

2. **Service Robotics:** Service robots are designed to assist humans in various tasks outside the industrial setting. Examples include healthcare robots that aid in surgery, robotic vacuum cleaners, and delivery drones.

3. **Autonomous Systems:** Autonomous robots and vehicles are capable of navigating and interacting with their environment without human intervention. This includes self-driving cars, autonomous drones, and even robots used in planetary exploration.

4. **Humanoid Robots:** These robots are designed to resemble humans in appearance and, to some extent, behavior. They are used in research on human-robot interaction, entertainment, and even as companions for the elderly or people with special needs.

5. **AI-Powered Robotics:** Robotics is increasingly intertwined with artificial intelligence. Robots are equipped with machine learning and deep learning algorithms, enabling them to adapt, learn, and make decisions based on sensory data.

6. **Extreme Environments:** Robots are employed in environments too hazardous or inaccessible for humans, such as deep-sea exploration, nuclear reactor maintenance, and space missions.

7. **Consumer Electronics:** Everyday items like smartphones and gaming consoles often incorporate robotics technology, from the sensors that enable screen rotation to voice-activated virtual assistants.

8. **Agricultural Automation:** In agriculture, robots are used for tasks like harvesting crops, monitoring soil conditions, and even autonomous tractors.

9. **Space Exploration:** Space robotics includes the design of robotic arms and rovers used in planetary exploration missions.

10. **Home Automation:** The integration of robotics into home environments for tasks like security, cleaning, and entertainment.

The scope of robotics continues to evolve as technology advances, creating new opportunities and challenges. Robotics is not just about building mechanical machines; it's about designing intelligent systems that can perceive, reason, and act in the world, ultimately enhancing our lives and transforming industries. As we delve deeper into this field, we'll explore the historical perspective of robotics and the key concepts that form its foundation.

B. Historical Perspective of Robotics and Automation

To fully grasp the significance and evolution of robotics and automation, one must journey through their rich historical timeline. The roots of these fields stretch back centuries, and their development has been marked by key inventions, discoveries, and innovations. Understanding this history provides valuable insights into the growth and potential of robotics and automation in contemporary society.

Ancient Automata: The concept of automation and mechanical devices has ancient origins. In ancient Greece, inventors like Archytas of Tarentum and Hero of Alexandria crafted automated devices, including steam-powered engines and automata that could perform simple tasks. Hero's "Pneumatica" and "Automata" treatises, written in the 1st century AD, are notable early examples of instructions for creating machines.

The Renaissance and Automata: During the Renaissance period, inventors and artisans continued to develop automata, often driven by clockwork mechanisms. Leonardo da Vinci, the quintessential Renaissance polymath, sketched designs for humanoid robots and mechanical knights. These endeavors laid the groundwork for future advancements.

The Industrial Revolution: The Industrial Revolution of the 18th and 19th centuries brought about a seismic shift in manufacturing and automation. Innovations like the power loom, steam engine, and assembly line revolutionized production processes. It was during this era that the term "automation" was first coined to describe the automatic control of machinery.

Early Robotics: The 20th century witnessed significant strides in robotics. In 1921, Czech playwright Karel Čapek introduced the term "robot" in his play "R.U.R. (Rossum's Universal Robots)," where robots were portrayed as artificial, labor-driven beings. This play popularized the term, and it soon became synonymous with mechanical devices capable of performing tasks autonomously.

The Unimate and Industrial Robotics: The late 1950s saw the birth of the first industrial robot, named Unimate. Invented by George Devol and Joseph Engelberger, the Unimate was installed at a General Motors plant in 1961, marking the practical use of robots in manufacturing. These early industrial robots performed tasks like die casting and welding, laying the foundation for the modern robotics industry.

Space Age and Robotics: The space race of the mid-20th century led to the development of robotic arms used in space exploration. The famous Canadarm, developed by the Canadian company SPAR Aerospace, was employed on NASA's Space

Shuttle missions, demonstrating the versatility of robotic systems.

The Rise of Personal Computers and AI: The proliferation of personal computers in the 1970s and 1980s provided the computational power needed for advancements in robotics and artificial intelligence. Researchers began developing robots with the ability to perceive and respond to their environment using sensors and early forms of AI.

Contemporary Robotics and Automation: Today, robotics and automation permeate almost every facet of our lives, from manufacturing and healthcare to transportation and entertainment. Robotics research has produced humanoid robots, autonomous vehicles, and AI-powered systems that continue to redefine what is possible.

As we journey through this historical perspective, it becomes evident that robotics and automation are not merely fields of science and engineering but a testament to human ingenuity, curiosity, and the relentless pursuit of progress. With each chapter in history, we have moved closer to the vision of machines seamlessly interacting with the world, shaping the way we live and work. This historical context sets the stage for the exploration of key concepts, applications, and challenges in the exciting world of robotics and automation.

C. Key Concepts in Robotics: Sensors, Actuators, and Control Systems

The foundational pillars of robotics are the triumvirate of sensors, actuators, and control systems. These essential elements enable machines to perceive their environment, make decisions, and execute actions with precision and intelligence. Understanding these key concepts is pivotal in mastering the field of robotics and automation.

1. Sensors: Sensors serve as the sensory organs of robots, allowing them to gather data about their surroundings. They play a critical role in providing information about the environment, objects, and the robot's own state. Here are some common types of sensors used in robotics:

- **Cameras:** Visual sensors capture images and videos, enabling robots to perceive objects, navigate, and recognize patterns and faces.

- **Lidar (Light Detection and Ranging):** Lidar sensors use lasers to measure distances and create 3D maps of the environment, crucial for navigation and obstacle avoidance.

- **IMUs (Inertial Measurement Units):** IMUs contain accelerometers and gyroscopes to measure the robot's orientation and motion, aiding in balance and navigation.

- **Ultrasonic Sensors:** These sensors emit high-frequency

sound waves and measure the time it takes for them to bounce back, allowing robots to estimate distances to objects.

- **Proximity Sensors:** Proximity sensors detect the presence or absence of nearby objects, used for collision avoidance and object detection.

- **Force and Tactile Sensors:** These sensors measure forces and pressure, enabling robots to handle delicate objects or apply controlled force in interactions.

2. Actuators: Actuators are the muscles of a robot, responsible for converting electrical or pneumatic signals into physical movements. They allow robots to interact with their environment and perform tasks. Key types of actuators include:

- **Electric Motors:** These include servos and stepper motors, used for precise control of robot joints and movements.

- **Pneumatic Actuators:** Air-driven actuators, such as pneumatic cylinders, are used for tasks requiring rapid motion or force.

- **Hydraulic Actuators:** These are used in heavy-duty applications requiring high force, such as construction equipment and industrial machines.

- **Muscle-Like Actuators:** Soft and compliant actuators, inspired by biological muscles, are used in soft robotics for safer human-robot interaction.

3. Control Systems: Control systems serve as the robot's brain, orchestrating the interaction between sensors and actuators to achieve desired tasks. They encompass various levels of control:

- **Low-Level Control:** This level focuses on real-time, low-level tasks such as motor control and sensor integration. It ensures that the robot responds to sensory input with precision.

- **Motion Planning:** At this level, robots determine how to move efficiently from one point to another, avoiding obstacles and following predefined trajectories.

- **Perception and Sensing:** Control systems process data from sensors, recognizing objects, estimating distances, and building a model of the environment.

- **Higher-Level Decision Making:** Advanced control systems use artificial intelligence and machine learning to make complex decisions. This includes path planning, obstacle avoidance, and even learning from experience.

- **Feedback Control:** Feedback loops continuously adjust the robot's actions based on the difference between desired and actual states, ensuring accuracy and stability.

The synergy of sensors, actuators, and control systems is the essence of robotics and automation. These components work in harmony to enable robots to perform tasks ranging from manufacturing and exploration to healthcare and service industries. As we delve deeper into the world of robotics, we'll explore how these key concepts are applied and integrated to create increasingly sophisticated and capable robotic systems.

D. Robotics and Automation Applications: Transforming Industries and Enhancing Lives

The vast and versatile landscape of robotics and automation applications is reshaping industries and making a profound impact on our daily lives. From manufacturing to healthcare, agriculture to space exploration, these technologies are revolutionizing the way we work, live, and interact with the world.

1. Industrial Robotics and Automation:

- *Manufacturing:* Industrial robots are central to modern manufacturing, performing tasks like welding, assembly, and material handling with precision and efficiency.

- *Quality Control:* Automated systems use vision sensors and AI to inspect products for defects, ensuring high-quality output.

- *Logistics and Warehousing:* Automated guided vehicles (AGVs) and robotic arms streamline material transport and order fulfillment in warehouses.

2. Healthcare Robotics:

- *Surgery:* Robotic surgical systems provide surgeons with enhanced precision and minimally invasive capabilities, improving patient outcomes.

- *Rehabilitation:* Assistive robots help patients recover from injuries or surgeries, offering therapeutic exercises and mobility support.

- *Telemedicine:* Robots enable remote consultations and diagnostics, enhancing access to healthcare services.

3. Agricultural Robotics:

- *Precision Agriculture:* Drones, autonomous tractors, and robotic harvesters optimize farming practices, increasing crop yield and reducing resource usage.

- *Weed and Pest Control:* Robots equipped with sensors and cameras identify and manage weeds and pests, reducing the need for pesticides.

4. Space Robotics:

- *Planetary Exploration:* Robotic rovers like NASA's Curiosity and Perseverance explore distant planets, conducting experiments and sending valuable data back to Earth.

- *Satellite Servicing:* Space robots can repair, refuel, or reposition satellites in orbit, extending their operational lifespan.

5. Service Robots:

- *Home Assistance:* Robot vacuum cleaners, smart speakers, and companion robots assist with household tasks, entertainment, and companionship.

- *Retail:* Retail robots can provide customer assistance, inventory management, and even cashier functions.

- *Hospitality:* Robots in hotels and restaurants offer services such as room delivery, cleaning, and food preparation.

6. Autonomous Vehicles and Drones:

- *Self-Driving Cars:* Autonomous vehicles use sensors, AI, and mapping technology to navigate roads safely and efficiently.

- *Delivery Drones:* Drones deliver packages, medicines, and essential supplies to remote or inaccessible areas.

- *Aerial Surveillance:* Drones are employed for surveillance, search and rescue, and environmental monitoring.

7. Robotics in Education and Research:

- *Educational Robots:* Robots designed for educational

purposes help students learn programming, engineering, and problem-solving.

- *Research Assistants:* Robots assist researchers in laboratories, handling repetitive tasks and data collection.

8. Environmental Monitoring:

- *Underwater Robotics:* Submersible robots explore the depths of the ocean, gathering data on marine life and geological formations.

- *Environmental Sensors:* Autonomous robots equipped with sensors monitor pollution, climate, and wildlife in remote environments.

9. Home Automation Systems:

- *Smart Homes:* Automation systems control lighting, heating, security, and appliances, enhancing convenience and energy efficiency.

10. IoT Automation and Robotics:

- *Internet of Things (IoT):* Robotics and automation are integrated into the IoT ecosystem, enabling smart cities, industrial IoT, and interconnected devices.

11. Entertainment and Gaming:

- *Entertainment Robots:* Robots are used in theme parks, movies, and interactive exhibits to entertain and engage audiences.

- *Gaming Industry:* Gaming platforms incorporate robotics and augmented reality (AR) for immersive gaming experiences.

The applications of robotics and automation are continuously expanding, driven by technological advancements and innovative solutions. These technologies have the potential to address critical challenges, improve efficiency, enhance safety, and enrich our lives in countless ways. As we delve deeper into this realm, we uncover the innovations and breakthroughs that are shaping the future of robotics and automation across industries.

E. Benefits and Challenges of Robotics and Automation

The integration of robotics and automation into various industries and aspects of our lives brings with it a host of advantages and, simultaneously, a set of complex challenges. Understanding these benefits and challenges is essential for harnessing the full potential of these technologies while mitigating potential drawbacks.

Benefits of Robotics and Automation:

1. **Enhanced Efficiency and Productivity:** One of the primary advantages is the significant increase in efficiency and productivity. Robots and automated systems can perform tasks with precision and speed 24/7, reducing production time and costs.

2. **Improved Quality:** Automation ensures consistent product quality by eliminating human errors and variations in manufacturing and assembly processes. This is particularly crucial in industries like automotive and electronics.

3. **Safety Improvement:** Robots can handle dangerous and repetitive tasks in hazardous environments, reducing the risk of injuries to human workers. They can also work in extreme conditions, such as deep-sea exploration or outer space missions, where human presence is impractical or perilous.

4. **Cost Reduction:** While there is an initial investment in robotics and automation, they can lead to long-term cost savings through reduced labor costs, lower error rates, and decreased waste in manufacturing.

5. **Increased Precision:** Robots can execute highly precise movements and measurements, making them invaluable in fields like healthcare, where precision is critical in surgeries and diagnostics.

6. **Flexibility and Scalability:** Automation systems can be easily reprogrammed and adapted to handle different tasks and products, making them versatile and suitable for evolving production needs.

7. **Data Collection and Analysis:** Automation systems equipped with sensors can collect vast amounts of data, enabling real-time monitoring and data-driven decision-making. This is particularly beneficial in quality control and predictive maintenance.

8. **Space Exploration:** Robots and automated systems play a vital role in space exploration, conducting experiments, gathering data, and even assisting in constructing structures on other planets.

9. **Labor Augmentation:** Collaborative robots, or cobots, work alongside humans, enhancing human productivity and job satisfaction by handling physically strenuous or repetitive tasks.

10. **Environmental Benefits:** Automation can lead to energy savings and reduced waste, contributing to sustainability goals. For example, automation in agriculture can optimize resource usage.

Challenges of Robotics and Automation:

1. **High Initial Costs:** Implementing robotics and automation systems often requires a substantial upfront investment in equipment, training, and infrastructure, which can be a barrier for smaller businesses.

2. **Job Displacement:** The fear of job loss due to automation is a significant concern. While automation can create new job opportunities, it can also render some roles obsolete, requiring retraining and workforce transition.

3. **Technical Complexity:** Designing, programming, and maintaining robotic systems can be complex and requires specialized skills and expertise. This can pose challenges for businesses seeking to adopt automation.

4. **Security Risks:** As systems become more interconnected, they are vulnerable to cybersecurity threats. Protecting sensitive data and ensuring the security of automation systems is a constant challenge.

5. **Ethical Dilemmas:** With the integration of AI and robotics, ethical concerns arise, such as AI bias, privacy issues, and the potential for autonomous weapons.

6. **Human-Robot Interaction:** Ensuring safe and effective collaboration between humans and robots, especially in

dynamic and unstructured environments, presents technical challenges.

7. **Regulatory Compliance:** Adhering to safety and industry regulations is crucial but can be complex, particularly for sectors like healthcare and autonomous vehicles.

8. **Maintenance and Downtime:** Robots and automated systems require regular maintenance, and downtime during maintenance can impact productivity.

9. **Resistance to Change:** Cultural and organizational resistance to automation can hinder its successful adoption. Employees may be hesitant to embrace automation due to job security concerns or fear of job displacement.

10. **Environmental Concerns:** The energy consumption of automated systems and the disposal of obsolete robotic components can have environmental implications.

Balancing the benefits and challenges of robotics and automation requires a thoughtful and strategic approach. When implemented thoughtfully and with consideration of these factors, robotics and automation have the potential to drive innovation, improve quality of life, and fuel economic growth.

CHAPTER 2

Robotics Fundamentals: Exploring the Building Blocks of Automation

In the captivating world of robotics, understanding the fundamentals is akin to deciphering the code to an extraordinary universe of possibilities. Robotics fundamentals lay the groundwork for comprehending the core principles, essential components, and intricate mechanics that breathe life into robots. As we embark on this journey, we'll delve into the heart of robotics, uncovering the types of robots, dissecting their components, exploring kinematics and dynamics, and unraveling the intricacies of robot control systems. These fundamentals form the cornerstone of a comprehensive understanding of the captivating realm of robotics, where machines come alive, bridging the gap between imagination and reality. Join us as we delve into the essential elements that make robotics an art and science of boundless innovation and invention.

A. Types of Robots: Industrial, Service, and Autonomous

Robots, the versatile workhorses of the modern age, come in various forms tailored to their specific functions and environments. The categorization of robots into types provides insights into their design, capabilities, and applications. Three prominent categories are industrial robots, service robots, and autonomous robots, each serving distinct roles and contributing to the advancement of various industries.

1. Industrial Robots:

Overview: Industrial robots are synonymous with precision, efficiency, and automation in manufacturing and industrial settings. They excel in repetitive and high-precision tasks, making them indispensable in industries where consistent quality and high production rates are essential.

Key Characteristics:

- **Repetitive Tasks:** Industrial robots are designed to perform repetitive tasks, such as welding, painting, assembly, and material handling, with unerring precision.

- **Programmability:** They are programmable machines, allowing for flexibility in reconfiguring their operations to suit different production needs.

- **Accuracy:** These robots exhibit high precision, making them ideal for tasks demanding exact measurements.

- **Safety Measures:** Safety features like fencing, sensors, and interlocking mechanisms ensure the safety of human workers in their vicinity.

Applications:

- **Automotive Industry:** Industrial robots dominate automotive assembly lines, performing tasks such as welding, painting, and installing components.

- **Electronics Manufacturing:** They are employed in electronic assembly for soldering, precision placement, and circuit board testing.

- **Metal Fabrication:** Industrial robots cut, bend, and weld metal components in industries like aerospace and shipbuilding.

- **Packaging and Material Handling:** They play a crucial role in sorting, packaging, and palletizing products in warehouses and logistics centers.

2. Service Robots:

Overview: Service robots are designed to interact with and assist humans, whether in homes, healthcare facilities, or public spaces. They offer a wide range of services, from household chores to medical care and customer service.

Key Characteristics:

- **Human Interaction:** Service robots are intended to interact with humans in a friendly and non-threatening manner.

- **Versatility:** They have versatile applications, including cleaning, caregiving, reception, and entertainment.

- **Mobility:** Service robots may be mobile, navigating autonomously or semi-autonomously to reach their destinations.

- **AI and Sensors:** Many service robots incorporate artificial intelligence and sensors for tasks like facial recognition and object detection.

Applications:

- **Domestic Robots:** Vacuum cleaning robots (e.g., Roomba), robot pets, and smart speakers are part of the burgeoning market of home service robots.

- **Healthcare Robots:** Assistive robots help with tasks like lifting patients, delivering medications, and providing companionship to the elderly.

- **Hospitality and Retail:** Robots are used as concierges, greeters, and order assistants in hotels, restaurants, and stores.

- **Education:** Educational robots teach programming, language skills, and STEM concepts to students.

3. Autonomous Robots:

Overview: Autonomous robots are designed to operate independently, making decisions based on sensor inputs and artificial intelligence algorithms. They navigate, adapt to their environment, and perform tasks with minimal human intervention.

Key Characteristics:

- **Autonomy:** Autonomous robots can operate without continuous human control or guidance.

- **Sensors:** They are equipped with various sensors (e.g., lidar, cameras, IMUs) to perceive and interpret their surroundings.

- **Artificial Intelligence:** AI algorithms enable autonomous decision-making, path planning, and learning from experience.

- **Adaptability:** These robots can adapt to dynamic and unstructured environments, such as self-driving cars navigating city streets.

Applications:

- **Autonomous Vehicles:** Self-driving cars and drones are prime examples of autonomous robots revolutionizing transportation and logistics.

- **Agricultural Robotics:** Autonomous tractors and drones

optimize crop management and monitoring.

- **Space Exploration:** Rovers like NASA's Curiosity navigate distant planets autonomously, conducting experiments and sending data.

- **Search and Rescue:** Autonomous robots aid in locating survivors in disaster-stricken areas.

Understanding these categories of robots illuminates the diverse roles they play in modern society. Industrial robots drive manufacturing efficiency, service robots enhance our daily lives, and autonomous robots push the boundaries of what is achievable in exploration and automation. As technology advances, these categories continue to evolve, and new forms of robots emerge to tackle novel challenges and opportunities.

B. Robot Components: Manipulators, End Effectors, and Drives

The anatomy of a robot is a carefully orchestrated symphony of components, each serving a distinct role in enabling the machine to manipulate its environment and perform tasks. Among the key components are manipulators, end effectors, and drives, which collectively give a robot its remarkable capabilities.

1. Manipulators:

Overview: Manipulators are the mechanical arms or limbs of a robot. They provide the robot with mobility and dexterity, allowing it to move in space and interact with objects in its environment. Manipulators come in various forms, each designed for specific applications.

Key Characteristics:

- **Degrees of Freedom (DoF):** The number of independent movements a manipulator can make is referred to as its DoF. More DoF generally provide greater flexibility and dexterity.

- **Joints:** Manipulators consist of joints that allow rotational or translational movements. Common types include revolute (rotational) and prismatic (translational) joints.

- **End-Effector Mounting:** Manipulators have mounting points at their ends where various end effectors or tools can be attached.

Types of Manipulators:

- **Serial Manipulators:** These are the most common type, with a chain of connected links and joints. They are versatile and used in various applications.

- **Parallel Manipulators:** These have multiple arms connected to both a fixed and moving platform. They offer high precision and stiffness and are often used in CNC machines.

- **SCARA Robots (Selective Compliance Assembly Robot Arm):** SCARA robots have two parallel rotational joints and are known for their speed and precision. They are commonly used in assembly tasks.

- **Articulated Robots:** With multiple revolute joints, articulated robots mimic human arms and are used in applications requiring complex motions, like welding and painting.

2. End Effectors:

Overview: End effectors, also known as robot tools or grippers, are the attachments at the end of a robot's manipulator. They enable the robot to interact with and manipulate objects, perform tasks, and carry out specific functions. End effectors are highly specialized, tailored to the task at hand.

Key Characteristics:

- **Design:** End effectors can have various designs, including fingers for grasping, vacuum cups for lifting, welding torches for joining, and cutting tools for machining.

- **Actuation:** Some end effectors have active mechanisms, such as servo-driven grippers, while others are passive and rely on the robot's movements for operation.

- **Sensing:** Advanced end effectors may incorporate sensors like force/torque sensors or cameras to provide feedback during manipulation.

Types of End Effectors:

- **Grippers:** Grippers come in two-finger, three-finger, or multi-finger designs and are used for picking up and holding objects of various shapes and sizes.

- **Vacuum Cups:** These are used for handling flat, non-porous objects like sheets of glass or metal.

- **Welding Tools:** End effectors designed for welding tasks are equipped with welding torches and sensors for precise welding operations.

- **Cutting Tools:** End effectors for machining tasks may include cutting tools, drills, or milling heads.

3. Drives:

Overview: Drives are the mechanisms that generate motion and actuate the joints of a robot's manipulator. They provide the necessary power to move the robot's limbs, enabling it to execute tasks with precision and speed.

Key Characteristics:

- **Types:** Drives can be electric, hydraulic, or pneumatic, depending on the specific requirements of the robot and its application.

- **Control:** Drives are controlled by the robot's central processing unit (CPU) or controller, which sends signals to the drives to move the robot's joints.

- **Precision and Speed:** The choice of drives influences the robot's precision, speed, and load-carrying capacity.

Types of Drives:

- **Electric Drives:** Electric motors are commonly used for their precision and ability to provide precise control of joint movements. They are found in most modern industrial robots.

- **Hydraulic Drives:** Hydraulic systems use pressurized fluid to actuate joints, providing high force and power. They are used in heavy-duty applications like construction equipment.

- **Pneumatic Drives:** Pneumatic drives use compressed air to move joints, offering speed and simplicity. They are often found in lightweight robots and automation systems.

These components, working in harmony, transform a robot from a mere machine into a versatile tool capable of performing a wide range of tasks. Whether it's the precise articulation of manipulators, the specialized capabilities of end effectors, or the power of drives, each element plays a pivotal role in shaping the robot's functionality and defining its suitability for various applications.

C. Kinematics and Dynamics in Robotics

In the world of robotics, understanding kinematics and dynamics is akin to deciphering the language of motion and forces that govern the behavior of robots. These two fundamental concepts are essential for designing, controlling, and optimizing the movement and performance of robotic systems. Let's delve into the intricacies of kinematics and dynamics in robotics.

Kinematics:

1. Forward Kinematics:

- *Definition:* Forward kinematics is the study of how the joint angles and lengths of a robot's manipulator relate to the position and orientation of its end effector (the tool or gripper at the end of the robot arm).

- *Equations:* Mathematical equations, often represented using transformation matrices, are used to calculate the position and orientation of the end effector based on the joint angles.

- *Applications:* Forward kinematics is crucial for path planning, motion control, and visual simulation of a robot's movement.

2. Inverse Kinematics:

- *Definition:* Inverse kinematics is the reverse process. It involves determining the joint angles required to achieve a specific position and orientation of the end effector.

- *Challenges:* Solving inverse kinematics can be complex, as multiple solutions may exist, and some configurations may be physically impossible.

- *Applications:* Inverse kinematics is essential for tasks where precise positioning of the robot's end effector is required, such as pick-and-place operations or surgical procedures.

Dynamics:

1. Kinetics and Newton's Laws:

- *Kinetics:* Kinetics deals with the study of forces and torques (rotational forces) acting on a robot.

- *Newton's Laws:* The laws of motion formulated by Sir Isaac Newton are fundamental in robot dynamics. They describe how forces and torques affect the motion of a robot.

- *Applications:* Understanding kinetics and Newton's laws is crucial for designing robot structures that can withstand forces and for predicting how robots will move in response to applied forces.

2. Robot Dynamics Equations:

- *Equations of Motion:* Robot dynamics equations describe how the forces and torques applied to a robot's joints result in accelerations of those joints and the robot's overall motion.

- *Inverse Dynamics:* Inverse dynamics involves calculating the required forces and torques at the robot's joints to achieve a desired motion or trajectory.

- *Control:** Robot control systems use dynamic models to control the motion and stability of the robot in real-time, enabling tasks like walking for humanoid robots or precise control in manufacturing.

3. Gravitational and Inertial Forces:

- *Gravitational Forces:* Understanding how gravitational forces affect robot dynamics is crucial for ensuring stability, especially in bipedal or legged robots.

- *Inertial Forces:** The mass distribution of a robot affects its dynamics. Inertial forces come into play when a robot accelerates or decelerates.

- *Applications:** Gravitational and inertial forces are critical in applications like robotic arms, walking robots, and drones.

4. External Forces and Friction:

- *External Forces:** Robots often interact with external forces, such as contact with objects, obstacles, or environmental forces like wind or water.

- *Friction:** Frictional forces between a robot and its environment can significantly affect its motion.

- *Applications:** Understanding how robots interact with their environment is essential for tasks like manipulation, navigation, and stability.

Applications of Kinematics and Dynamics in Robotics:

- **Manufacturing:** Robot arms in manufacturing use kinematics for precise positioning, and dynamics for optimizing speed and accuracy.

- **Aerospace:** Kinematics and dynamics are essential in the design of robotic spacecraft arms and drones.

- **Medical Robotics:** Surgical robots rely on kinematics for precise positioning and dynamics for stability during procedures.

- **Autonomous Vehicles:** Self-driving cars and drones use dynamic models to navigate and respond to external forces.

- **Animation and Simulations:** Kinematics and dynamics are fundamental in computer graphics, animations, and video games.

- **Research:** Robotics researchers use these concepts to design and control robots for various applications, from exploration to disaster response.

In summary, kinematics and dynamics serve as the mathematical and physical foundations of robotics, enabling engineers and researchers to design, control, and optimize robots for a wide range of tasks and applications. These concepts are central to unlocking the full potential of robotics in our modern world.

D. Robot Control Systems: Orchestrating Precision and Autonomy

Robot control systems are the brains behind the brawn of robotic machines, responsible for orchestrating their movements, decision-making processes, and responses to external stimuli. These systems are at the heart of a robot's functionality, enabling it to perform tasks with precision, autonomy, and adaptability. Let's delve into the intricacies of robot control systems and their vital role in the world of robotics.

1. Central Processing Unit (CPU):

- *Definition:* The CPU is the computational core of a robot's control system, functioning as its "brain." It processes sensory data, executes algorithms, and sends commands to various robot components.

- *Processing Power:* The CPU's processing power dictates the robot's computational capabilities, affecting its ability to handle complex tasks and real-time control.

- *Control Algorithms:* Control algorithms, such as inverse kinematics and path planning, are executed by the CPU to translate high-level commands into low-level motor commands for the robot's actuators.

2. Sensory Input:

- *Sensor Types:* Sensors provide the robot with information about its environment, including cameras for vision, lidar for distance measurement, IMUs for orientation, and various other sensors for touch, pressure, and temperature.

- *Data Fusion:* Sensor data from multiple sources is fused and processed to create a comprehensive understanding of the robot's surroundings.

- *Feedback Loops:* Sensor feedback is essential for closed-loop control, where the robot adjusts its actions based on real-time data, ensuring accuracy and adaptability.

3. Actuator Control:

- *Actuators:* Actuators, such as motors, are responsible for moving the robot's joints and end effectors. The control system sends commands to actuators to achieve desired movements.

- *Real-Time Control:* Actuator control requires real-time responsiveness to maintain stability and precision during motion.

- *Motion Planning:* Control systems generate motion trajectories and commands for the actuators to follow, allowing the robot to move smoothly and accurately.

4. Control Paradigms:

- *Open-Loop Control:* In open-loop control, predefined commands are sent to actuators without feedback. While simple, it lacks adaptability to changing conditions.

- *Closed-Loop Control:* Closed-loop control incorporates sensor feedback to continuously adjust commands, enabling the robot to respond to external disturbances and achieve desired outcomes.

- *Feedback Control:** PID (Proportional-Integral-Derivative) controllers are commonly used in feedback control systems to regulate variables like position and velocity.

5. Motion Control:

- *Trajectory Planning:* Motion control involves planning and executing desired trajectories for the robot's end effector or joints. This includes path planning for navigation and motion profiles for smooth movements.

- *Obstacle Avoidance:* Control systems incorporate collision detection and avoidance algorithms to ensure the robot can navigate safely in its environment.

6. Behavior-Based Robotics:

- *Behavior-Based Systems:* Some robots employ behavior-based control systems, where behaviors or modules responsible for specific tasks or reactions are combined to create complex behaviors.

- *Hierarchical Control:** In hierarchical control, different layers or levels of behaviors are organized to enable robots to handle high-level goals while managing lower-level tasks.

7. Robot Simulation and Testing:

- *Simulation:** Before deploying robots in real-world scenarios, control systems are often tested in simulated environments to validate their functionality and performance.

- *Training and Learning:** Machine learning and reinforcement learning algorithms can be integrated into control systems, allowing robots to adapt and improve their performance over time.

8. Real-Time Control and Safety:

- *Real-Time Operating Systems (RTOS):* Many robots require real-time control to ensure timely responses. RTOSs manage the scheduling of tasks and processes for real-time applications.

- *Safety Measures:* Control systems implement safety features such as emergency stop buttons, collision detection, and torque limits to protect humans and the robot itself.

Applications of Robot Control Systems:

- **Manufacturing:** Industrial robots use control systems for precise and repetitive tasks in manufacturing.

- **Autonomous Vehicles:** Control systems are pivotal in self-driving cars and drones for navigation and collision avoidance.

- **Healthcare:** Surgical robots rely on control systems for precise movements during medical procedures.

- **Agriculture:** Agricultural robots use control systems for tasks like autonomous harvesting and crop monitoring.

- **Space Exploration:** Robotic rovers and spacecraft depend on control systems for navigation and operation in space.

In essence, robot control systems are the architects of robotic capabilities, enabling machines to interact with the physical world, execute tasks with precision, and adapt to changing conditions. As technology continues to advance, control systems play a central role in shaping the capabilities and possibilities of robotics in various industries and applications.

CHAPTER 3

Sensors, Perception, and Automation Technologies: The Senses of the Robotic World

In the realm of robotics and automation, the ability to sense, perceive, and react to the environment is akin to the senses that guide human beings. Sensors, perception systems, and automation technologies serve as the eyes, ears, and intellect of robotic entities, endowing them with the power to interact, navigate, and make informed decisions. This journey explores the multifaceted world of sensors and perception, unveiling the technologies that bridge the gap between the physical and the digital, ultimately reshaping industries, enhancing safety, and advancing the frontiers of automation. Join us as we delve into the marvels of sensory perception and automation technologies, where the fusion of data and intelligence paves the way for a smarter, more connected future.

A. Sensors in Robotics and Automation: Visionaries of the Robotic World

Sensors are the vigilant eyes and ears of robots and automation systems, allowing them to perceive and interact with their surroundings. In the dynamic world of robotics and automation, sensors play a pivotal role in collecting data, making decisions, and executing tasks with precision. This in-depth exploration focuses on three crucial sensor types in robotics and automation: cameras, LiDAR (Light Detection and Ranging), and IMUs (Inertial Measurement Units).

1. Cameras:

Overview: Cameras are among the most versatile and widely used sensors in robotics and automation. They capture visual information, allowing robots to analyze their environment, recognize objects, and make informed decisions based on images and videos.

Types of Cameras:

- **RGB Cameras:** These capture color images and are commonly used for tasks like object recognition, tracking, and navigation.

- **Depth Cameras:** Depth-sensing cameras, such as Microsoft Kinect or LiDAR-equipped cameras, provide depth information, enabling robots to perceive the 3D structure of their surroundings.

- **Thermal Cameras:** Thermal imaging cameras detect heat signatures and are valuable in applications like search and rescue, where identifying warm bodies in low visibility conditions is crucial.

Applications of Cameras in Robotics:

- **Object Detection:** Cameras identify and locate objects in a robot's field of view, enabling tasks like pick-and-place in manufacturing.

- **Autonomous Navigation:** Cameras help robots navigate through environments, avoid obstacles, and follow predefined paths.

- **Visual Simultaneous Localization and Mapping (VSLAM):** Cameras are used in VSLAM systems to create maps of environments and determine a robot's position within them.

- **Surveillance and Security:** Cameras play a key role in monitoring and surveillance applications, including home security and industrial safety.

- **Medical Imaging:** In healthcare, cameras are used for procedures like endoscopy, enabling minimally invasive diagnostics and surgeries.

2. LiDAR (Light Detection and Ranging):

Overview: LiDAR is a remote sensing technology that measures distances and creates high-resolution 3D maps of the environment using laser pulses. LiDAR sensors emit laser beams and measure the time it takes for the light to bounce back, allowing for precise distance calculations.

Types of LiDAR Sensors:

- **Solid-State LiDAR:** These sensors use solid-state components like microelectromechanical systems (MEMS) to steer laser beams and are compact and durable.

- **Mechanical LiDAR:** Mechanical LiDAR sensors have rotating mirrors or spinning assemblies to scan the environment and are typically bulkier but offer high precision.

Applications of LiDAR in Robotics:

- **Autonomous Vehicles:** LiDAR is a crucial sensor in self-driving cars and autonomous drones for real-time mapping and obstacle avoidance.

- **Agriculture:** LiDAR-equipped drones are used in precision agriculture for crop monitoring and terrain mapping.

- **Surveying and Mapping:** LiDAR is employed in geospatial applications, such as topographic surveys and land management.

- **Urban Planning:** LiDAR helps city planners create detailed 3D models of urban environments for infrastructure development and disaster planning.

- **Forestry:** LiDAR is used to estimate tree height, canopy density, and other forest parameters for sustainable forestry management.

3. IMUs (Inertial Measurement Units):

Overview: IMUs are sensors that measure the linear and angular motion of a robot or object. They typically include accelerometers, gyroscopes, and sometimes magnetometers. IMUs provide crucial information about a robot's orientation and movement.

Applications of IMUs in Robotics:

- **Robot Localization:** IMUs aid in estimating a robot's position and orientation, especially in GPS-denied environments or when GPS signals are unreliable.

- **Motion Control:** IMUs are used for stabilizing and controlling robotic platforms like drones, balancing robots, and autonomous vehicles.

- **Gesture Recognition:** IMUs can be integrated into wearable devices and human-robot interfaces to recognize gestures and movements.

Integration and Fusion: In many robotics applications, these sensor types are not used in isolation. Sensor fusion techniques combine data from multiple sensors, such as cameras, LiDAR, and IMUs, to create a more comprehensive and accurate perception of the environment. This integration is particularly critical in autonomous navigation, object recognition, and robotics applications that require real-time decision-making.

In conclusion, cameras, LiDAR, and IMUs are essential sensory instruments that empower robots and automation systems to interact with and understand their surroundings. The synergy between these sensors enables robots to perform a wide array of tasks with precision, adaptability, and safety, propelling advancements in robotics and automation across various industries.

B. Object Detection and Localization in Robotics and Automation

Object detection and localization are fundamental capabilities that empower robots and automation systems to perceive their environment, identify objects of interest, and determine their precise locations. These capabilities are essential for a wide range of applications, from autonomous navigation to industrial automation, and play a pivotal role in enhancing the intelligence and functionality of robots. In this in-depth exploration, we delve into the intricacies of object detection and localization in the realm of robotics and automation.

Object Detection:

1. Overview:

- Object detection is the process of identifying and locating objects within an image or a scene captured by sensors, typically cameras or LiDAR, used in robots and automation systems.

- It involves not only recognizing objects but also determining their boundaries, often represented as bounding boxes.

2. Techniques for Object Detection:

- **Classical Computer Vision:** Traditional computer vision techniques, such as Haar cascades, Histogram of Oriented Gradients (HOG), and edge detection, were used for object detection in the past.

- **Deep Learning:** Modern object detection heavily relies on deep learning approaches, particularly Convolutional Neural Networks (CNNs). Models like Faster R-CNN, YOLO (You Only Look Once), and SSD (Single Shot MultiBox Detector) have revolutionized object detection by achieving impressive accuracy and real-time performance.

3. Challenges in Object Detection:

- **Variability:** Objects can appear in various sizes, orientations, and lighting conditions, making detection challenging.

- **Occlusion:** Objects may be partially or fully obscured by other objects or obstacles.

- **Real-Time Requirements:** Many applications, such as autonomous vehicles, demand real-time object detection, necessitating efficient algorithms and hardware.

4. Applications of Object Detection in Robotics:

- **Autonomous Navigation:** Self-driving cars, drones, and mobile robots use object detection to identify pedestrians, vehicles, and obstacles in their path.

- **Manufacturing and Quality Control:** Industrial robots employ object detection to locate and manipulate objects on assembly lines.

- **Agriculture:** Drones equipped with object detection technology monitor crop health and detect pests.

- **Healthcare:** Robots in healthcare settings can detect and locate medical instruments, patients, and obstacles in their environment.

Object Localization:

1. Overview:

- Object localization is the process of not only detecting an object but also determining its precise position within the sensor's field of view.

- It typically involves estimating the object's center or a key landmark within the object.

2. Techniques for Object Localization:

- **Bounding Box Regression:** Object detection models often predict bounding boxes around objects. Localization can be achieved by extracting the center coordinates or other landmarks from these boxes.

- **Keypoint Detection:** In some cases, object localization involves identifying specific keypoints or landmarks on the object, such as facial features in facial recognition.

3. Challenges in Object Localization:

- **Scale and Aspect Ratio:** Accurate localization requires handling objects of varying sizes and aspect ratios.

- **Complex Backgrounds:** Distinguishing objects from cluttered or textured backgrounds can be challenging.

- **Partial Occlusion:** Objects may be partially occluded, making precise localization difficult.

4. Applications of Object Localization in Robotics:

- **Robot Manipulation:** Robots need precise localization to grasp objects effectively and perform tasks like pick-and-place operations.

- **Augmented Reality (AR):** AR applications superimpose virtual objects onto the real world, necessitating accurate object localization.

- **Medical Imaging:** Object localization is crucial in medical imaging for pinpointing specific anatomical structures or anomalies.

- **Security and Surveillance:** Object localization enhances the tracking and monitoring of individuals and objects in surveillance systems.

Sensor Fusion and Multi-Object Tracking: In many robotics and automation applications, object detection and localization are not isolated tasks. Sensor fusion techniques combine data from multiple sensors, such as cameras, LiDAR, and IMUs, to create a more comprehensive perception of the environment. Multi-object tracking algorithms enable robots to not only detect and locate objects but also track their movements over time, enhancing their situational awareness and decision-making capabilities.

In conclusion, object detection and localization are foundational elements of robotic perception, enabling robots and automation systems to interact intelligently with their surroundings. These capabilities are essential for a wide array of applications across industries, from enhancing safety in autonomous vehicles to improving efficiency in manufacturing processes. As technology continues to advance, object detection and localization algorithms and systems will play an increasingly pivotal role in shaping the future of robotics and automation.

C. Environment Mapping in Robotics and Automation: Navigating the Unseen

Environment mapping is a critical capability that empowers robots and automation systems to understand and interact with their surroundings. It involves creating a digital representation of the physical world, allowing robots to navigate, plan paths, avoid obstacles, and make informed decisions. In this in-depth exploration, we delve into the intricacies of environment mapping, its various techniques, and its pivotal role in the realm of robotics and automation.

1. Types of Environment Mapping:

- **2D Grid Maps:** 2D grid maps represent the environment as a grid, where each cell can be marked as occupied or free. They are commonly used in applications like mobile robotics and indoor navigation.

- **3D Point Clouds:** 3D point clouds are collections of 3D points in space, often acquired using sensors like LiDAR or depth cameras. They provide detailed information about the environment's geometry.

- **Octree Maps:** Octrees are tree data structures that partition 3D space into hierarchical levels. They are used for efficient 3D mapping and are particularly useful for environments with varying levels of detail.

- **Semantic Maps:** Semantic maps go beyond geometry and incorporate semantic information about the environment, such as object labels or room classifications.

2. Mapping Techniques:

- **SLAM (Simultaneous Localization and Mapping):** SLAM is a fundamental technique that allows a robot to build a map of its environment while simultaneously estimating its own position within that map. It relies on sensor data, such as visual odometry, GPS, or LiDAR, and optimization algorithms.

- **Occupancy Grid Mapping:** This technique uses 2D grids to represent the environment, with each cell indicating the probability of occupancy. Probabilistic models, like Bayes filters, are often employed to update the map based on sensor measurements.

- **Volumetric Mapping:** Volumetric maps, such as Octrees or Truncated Signed Distance Fields (TSDF), represent 3D space as a set of voxels or points with associated occupancy probabilities or distance values. These maps are used for detailed 3D reconstruction.

- **Feature-Based Mapping:** Instead of representing the entire environment, feature-based mapping focuses on identifying and tracking distinctive features or landmarks, making it suitable for situations with limited computational resources.

3. Challenges in Environment Mapping:

- **Sensor Noise:** Sensors can introduce noise and uncertainty into measurements, requiring sophisticated filtering and fusion techniques for accurate mapping.

- **Dynamic Environments:** Mapping becomes challenging when the environment is dynamic, with objects moving or changing position.

- **Loop Closure:** Maintaining consistency in a map over time, especially in SLAM, requires the detection and closure of looped trajectories.

4. Applications of Environment Mapping in Robotics and Automation:

- **Autonomous Vehicles:** Self-driving cars and drones use environment mapping to navigate, detect obstacles, and plan safe paths.

- **Robotic Exploration:** Robots used in space exploration, search and rescue, or hazardous environments rely on mapping to plan and execute missions.

- **Industrial Automation:** In manufacturing and warehousing, robots use maps to optimize their movements and perform tasks efficiently.

- **Agriculture:** Agricultural robots use maps to plan routes, monitor crops, and perform precision tasks like planting and harvesting.

- **Construction:** In construction automation, robots can map construction sites to plan and execute tasks like bricklaying or concrete pouring.

Sensor Fusion and Localization: Environment mapping often goes hand in hand with sensor fusion and localization. Sensor fusion combines data from various sensors, such as cameras, LiDAR, IMUs, and GPS, to create a more accurate and comprehensive map. Localization techniques help robots determine their precise position within the map, enabling them to navigate and interact with the environment effectively.

In conclusion, environment mapping is a foundational capability that empowers robots and automation systems to understand, navigate, and interact with their surroundings. As technology continues to advance, the accuracy and efficiency of mapping techniques will play an increasingly pivotal role in reshaping industries and advancing the capabilities of robotics and automation.

D. Sensor Fusion: Merging Senses for Enhanced Perception

Sensor fusion is a pivotal technique in the field of robotics and automation, enabling systems to combine information from multiple sensors to create a more accurate, comprehensive, and reliable perception of their environment. This process mimics the way humans integrate information from various senses like vision, hearing, and touch to form a holistic understanding of the world. In this in-depth exploration, we delve into the intricacies of sensor fusion, its techniques, challenges, and the wide-ranging applications it empowers.

1. Types of Sensors in Sensor Fusion:

- **Cameras:** Visual sensors capture images and videos,

providing rich information about objects, shapes, and colors in the environment.

- **LiDAR (Light Detection and Ranging):** LiDAR sensors emit laser beams to measure distances and create 3D point clouds, offering precise spatial information.

- **IMUs (Inertial Measurement Units):** IMUs consist of accelerometers and gyroscopes, providing information about the robot's motion, orientation, and velocity.

- **GPS (Global Positioning System):** GPS provides absolute position and velocity information based on satellite signals.

- **Radar:** Radar sensors use radio waves to detect objects and measure their distances and velocities.

- **Ultrasonic Sensors:** Ultrasonic sensors use sound waves to detect nearby objects and measure distances.

- **Thermal Sensors:** Thermal sensors detect temperature variations in the environment and can be used for object detection or tracking warm objects.

2. Techniques of Sensor Fusion:

- **Data Level Fusion:** In data-level fusion, raw data from sensors are combined. This may involve combining images, point clouds, or sensor readings into a unified data stream.

- **Feature Level Fusion:** Feature-level fusion involves extracting relevant features from sensor data, such as object keypoints or edges, and then fusing these features to make decisions.

- **Decision Level Fusion:** In decision-level fusion, decisions or classifications made by individual sensors are combined to make a final decision. This is common in applications like voting systems.

- **State Estimation:** Sensor fusion is often used for state estimation, where the system estimates its own state (position, orientation, velocity) based on sensor data. Techniques like Kalman filters and particle filters are used for this purpose.

3. Challenges in Sensor Fusion:

- **Sensor Calibration:** Sensors may have different measurement units, noise characteristics, or alignment offsets, which must be calibrated for accurate fusion.

- **Temporal Alignment:** Sensor data may not be perfectly synchronized in time, which can lead to errors when fusing information.

- **Sensor Failures:** Sensors can fail or provide erroneous data. Robust sensor fusion algorithms must be capable of handling such scenarios.

- **Redundancy and Resource Usage:** Using multiple sensors increases computational and power requirements, which can be a challenge in resource-constrained systems.

4. Applications of Sensor Fusion:

- **Autonomous Vehicles:** Self-driving cars rely on sensor fusion to perceive their surroundings, detect obstacles, and make decisions about navigation and control.

- **Robotics:** Robots use sensor fusion for tasks such as mapping, localization, object recognition, and manipulation.

- **Augmented Reality (AR):** AR applications combine data from cameras, IMUs, and other sensors to overlay virtual information on the real world.

- **Defense and Surveillance:** Military and security systems use sensor fusion to enhance situational awareness and target tracking.

- **Healthcare:** Medical devices like MRI machines use sensor fusion to create detailed images by combining data from multiple sensors.

- **Environmental Monitoring:** Weather forecasting and environmental monitoring systems use data from various sensors to predict and track weather patterns.

Future Trends in Sensor Fusion:

Advancements in sensor technology, including the integration of AI and machine learning, are driving the development of more sophisticated sensor fusion techniques. These trends are expected to improve the robustness, accuracy, and real-time capabilities of sensor fusion systems, making them even more essential in the fields of robotics and automation.

In summary, sensor fusion is a cornerstone of modern robotics and automation, allowing systems to perceive and understand their environment with a higher degree of accuracy and reliability. As technology continues to evolve, sensor fusion will play an increasingly vital role in enabling robots and automated systems to interact with and navigate the complex, dynamic world around them.

E. Automation Technologies: Sensors, Actuators, and Control Systems

Automation technologies, including sensors, actuators, and control systems, constitute the critical infrastructure that enables robots and automated systems to perceive, interact with, and respond to their environments. In this comprehensive exploration, we delve into the fundamental components of automation, understanding how sensors gather data, actuators execute actions, and control systems orchestrate the entire process. These technologies form the bedrock of robotics and automation, driving innovation and practical applications across diverse fields.

1. Sensors:

Sensors in Robotics and Automation:

Sensors serve as the sensory organs of robots and automated systems, allowing them to perceive and interpret the world around them. These devices capture data from the environment, providing essential information for decision-making and control. Key sensor types include:

- **Cameras:** Vision systems, such as RGB and depth cameras, enable robots to capture visual data. Computer vision algorithms process this data for tasks like object recognition and navigation.

- **LiDAR (Light Detection and Ranging):** LiDAR sensors emit laser beams to create 3D maps of surroundings, enabling precise localization, mapping, and obstacle avoidance, crucial for autonomous vehicles and drones.

- **IMUs (Inertial Measurement Units):** IMUs consist of accelerometers and gyroscopes that measure acceleration and rotation. They are essential for stabilizing robots and tracking their orientation.

Sensors in Action:

Consider an autonomous delivery robot equipped with a LiDAR sensor. The LiDAR continuously scans its environment, creating a 3D map of obstacles. The robot's control system uses this data to plan safe navigation paths, avoiding collisions with objects and pedestrians.

2. Actuators:

Robot Actuators:

Actuators are the muscles of robots, responsible for executing physical actions. These devices convert electrical or pneumatic signals into mechanical motion. Common actuators in robotics include:

- **Electric Motors:** Servo motors and stepper motors provide precise control over movement, enabling robots to manipulate objects, traverse terrain, and perform tasks with accuracy.

- **Pneumatic Actuators:** Pneumatic systems use compressed air to drive motion, making them suitable for applications requiring speed and force, such as industrial automation.

- **Hydraulic Actuators:** Hydraulic systems use pressurized fluids to generate motion. They are employed in heavy-duty applications, like construction and manufacturing.

Actuators in Action:

Imagine a robotic arm equipped with electric motors. These motors control the arm's joints and grippers. As the control system sends commands to the motors, the arm precisely moves to pick up an object and place it in a designated location.

3. Control Systems:

Control Systems in Robotics and Automation:

Control systems are the brains behind automated operations. They receive data from sensors, process it, and send commands to actuators to achieve desired outcomes. Types of control systems include:

- **PID (Proportional-Integral-Derivative) Controllers:** PID controllers regulate continuous processes by adjusting control outputs based on error signals, providing stability and accuracy.

- **Feedback Control Systems:** These systems continuously measure outputs and adjust inputs based on feedback, ensuring that desired states or trajectories are maintained.

- **Machine Learning-Based Control:** Advanced control systems use machine learning algorithms to adapt to changing environments and learn from experience, enabling robots to improve their performance over time.

Control Systems in Action:

Consider a self-driving car navigating through traffic. The control system processes data from sensors, such as cameras and LiDAR, to identify obstacles, detect lane markings, and calculate safe speeds. It then sends commands to actuators, controlling the steering, acceleration, and braking to ensure the vehicle safely follows its route.

In conclusion, automation technologies, encompassing sensors, actuators, and control systems, are the lifeblood of robotics and automation. They enable robots and automated systems to sense, decide, and act in a wide range of applications, from manufacturing and healthcare to space exploration and transportation. As these technologies continue to advance, the potential for automation to transform industries and improve our quality of life remains boundless.

CHAPTER 4

Robot Manipulation, Mobility, and Process Automation: Mastering the Art of Robotic Interaction

In the world of robotics and automation, the ability to manipulate objects, move through diverse environments, and streamline complex processes is at the heart of technological innovation. Robot Manipulation, Mobility, and Process Automation represent the trifecta of capabilities that enable machines to seamlessly integrate into our dynamic world. From precise object manipulation on factory floors to autonomous navigation in unstructured terrain and the orchestration of intricate workflows, this journey explores the intricacies of these domains, revealing how robots have become indispensable partners in industry, healthcare, logistics, and beyond. Join us as we delve into the art of robotic interaction, where dexterity, mobility, and automation converge to redefine the boundaries of possibility.

A. Forward and Inverse Kinematics in Robotics: Decoding the Movement

Forward and Inverse Kinematics are fundamental concepts in the field of robotics that govern how robots move and interact with their environment. These mathematical techniques are crucial for understanding and controlling the motion of robotic manipulators, arms, and legs. In this in-depth exploration, we will delve into the intricacies of Forward and Inverse Kinematics, unraveling their roles, applications, and significance in the realm of robotics.

1. Forward Kinematics: Predicting End-Effector Position:

Overview: Forward Kinematics (FK) is the process of determining the position and orientation of a robot's end-effector (such as a gripper or tool) in relation to its joint angles and lengths. It calculates where the end-effector will be when specific joint values are applied.

Mathematical Representation: In a robotic manipulator with multiple joints, FK involves using the Denavit-Hartenberg parameters or transformation matrices to compute the position and orientation of the end-effector with respect to a reference frame.

Applications of Forward Kinematics:

- **Robot Control:** FK is essential for robot control and trajectory planning. It enables robots to reach desired positions and orientations in space.

- **Animation and Gaming:** In computer graphics and animation, FK is used to animate characters' movements and interactions with objects.

- **3D Printing:** FK is employed in 3D printing to calculate the path the printer head should follow to create a desired object.

2. Inverse Kinematics: Solving the Puzzle of Joint Angles:

Overview: Inverse Kinematics (IK) is the reverse process of Forward Kinematics. It involves finding the joint angles or parameters required to position the end-effector at a desired location and orientation.

Challenges in Inverse Kinematics:

- **Non-Linearity:** The relationship between joint angles and end-effector position is often nonlinear and complex.

- **Redundancy:** In multi-DOF robots, there can be multiple solutions to achieve the same end-effector position.

- **Singularity:** Certain configurations may lead to singularities, where IK solutions become unstable or degenerate.

Applications of Inverse Kinematics:

- **Robotics:** IK is essential for tasks like pick-and-place, robotic surgery, and manipulation of objects in unstructured environments.

- **Animation:** In computer graphics and character animation, IK is used to control the movement of limbs, fingers, and other body parts.

- **Simulations:** IK is applied in physics simulations to model the movement of articulated structures.

3. The Synergy of Forward and Inverse Kinematics:

Forward and Inverse Kinematics work together to enable precise control and coordination of robotic manipulators. While FK calculates the end-effector's position based on joint values, IK computes the joint values required to achieve a specific end-effector position.

4. Challenges and Considerations:

- **Singularity Avoidance:** When designing robotic systems, engineers often employ techniques to avoid or mitigate singularities to ensure stable and safe operations.

- **Real-Time Requirements:** In applications requiring real-time control, solving IK equations efficiently can be challenging.

5. Beyond Manipulators: While Forward and Inverse Kinematics are often associated with robotic arms, they are applicable to various robotic systems, including humanoid robots, mobile robots, and even biomechanical models used in medical simulations and exoskeletons.

In conclusion, Forward and Inverse Kinematics are pivotal concepts that govern the movement and manipulation of robots in a wide range of applications. Whether it's a manufacturing robot assembling intricate components or an animated character mimicking human movements, these mathematical techniques underpin the dexterity and versatility of modern robotics, continually pushing the boundaries of what robots can achieve.

B. Trajectory Planning in Robotics: Charting the Path to Precision

Trajectory planning is a critical aspect of robotics that enables machines to move from one point to another while navigating obstacles, avoiding collisions, and adhering to constraints. It plays a pivotal role in various applications, from autonomous vehicles and industrial robots to drones and humanoid robots. In this in-depth exploration, we will delve into the intricacies of trajectory planning, its methodologies, challenges, and the wide-ranging impact it has on the world of robotics.

1. What is Trajectory Planning?

Trajectory planning is the process of determining a feasible path for a robot or vehicle to follow while considering factors such as the robot's dynamics, environment, and task requirements. It involves calculating a sequence of positions, velocities, and accelerations over time to smoothly and safely reach the desired destination.

2. Key Components of Trajectory Planning:

- **Path Generation:** The first step involves defining the geometric path that the robot should follow. This path typically consists of a series of waypoints or key positions in the environment.

- **Velocity Profile:** Once the path is defined, the trajectory planner generates a velocity profile that specifies how fast the robot should move at each point along the path. This ensures smooth and controlled motion.

- **Collision Avoidance:** Trajectory planners incorporate collision avoidance strategies to ensure that the robot can navigate around obstacles or other dynamic elements in its environment.

- **Dynamic Constraints:** The planner considers the robot's physical limitations, such as maximum velocity, acceleration, and jerk, to generate trajectories that are within the robot's capabilities.

3. Trajectory Planning Methods:

- **Polynomial Trajectories:** These trajectories are defined by polynomial functions (e.g., quintic splines) and are well-suited for smooth, continuous paths.

- **Optimization-Based Methods:** These methods use

optimization algorithms to find trajectories that minimize a cost function while satisfying constraints. Model Predictive Control (MPC) is an example of this approach.

- **Sampling-Based Methods:** Algorithms like Rapidly-exploring Random Trees (RRT) and Probabilistic Roadmaps (PRM) sample the configuration space to find feasible paths through complex environments.

- **Machine Learning:** Deep reinforcement learning and neural network-based approaches have been applied to learn trajectory planning policies in certain applications.

4. Challenges in Trajectory Planning:

- **High-Dimensional Spaces:** Planning in high-dimensional configuration spaces, especially for robots with many degrees of freedom, can be computationally challenging.

- **Real-Time Constraints:** Many robotics applications, such as autonomous vehicles, require trajectory planning to be executed in real-time, imposing strict computational limitations.

- **Dynamic Environments:** Handling dynamic obstacles or changes in the environment adds complexity to trajectory planning.

- **Local vs. Global Planning:** Deciding when to replan the entire trajectory versus making local adjustments is a trade-off that planners must consider.

5. Applications of Trajectory Planning:

- **Autonomous Vehicles:** Trajectory planning is vital for self-driving cars, drones, and autonomous ground vehicles to navigate safely on roads and in urban environments.

- **Industrial Automation:** Robots in manufacturing plants use trajectory planning to move and manipulate objects with precision.

- **Aerospace:** Aircraft and spacecraft employ trajectory planning for navigation, trajectory correction, and orbital maneuvers.

- **Humanoid Robots:** Humanoid robots use trajectory planning to walk, run, or perform complex motions such as dancing.

- **Healthcare:** Surgical robots require precise trajectory planning for minimally invasive procedures.

- **Simulation and Gaming:** Trajectory planning is used in computer graphics and game development to control character movements and simulate physical interactions.

In conclusion, trajectory planning is a cornerstone of robotics that enables machines to move with precision and agility in a wide range of environments and applications. As technology advances, the development of more efficient and adaptive trajectory planning algorithms continues to drive progress in robotics, making automation safer, more capable, and more pervasive in our daily lives.

C. Robot Grasping and Manipulation: The Art of Precision and Dexterity

Robot grasping and manipulation represent the pinnacle of robotic capabilities, allowing machines to interact with the physical world, handle objects, and perform intricate tasks with precision and dexterity. These fundamental skills are essential in various applications, from manufacturing and logistics to healthcare and research. In this in-depth exploration, we will delve into the intricacies of robot grasping and manipulation, uncovering their methodologies, challenges, and transformative impact on the realm of robotics.

1. What is Robot Grasping and Manipulation?

Robot grasping and manipulation refer to the ability of robots to physically interact with objects in their environment. Grasping involves the process of gripping or holding an object, while manipulation includes a wide range of actions, such as lifting, moving, rotating, and placing objects. Together, these skills enable robots to perform tasks that require interaction with the physical world.

2. Key Components of Robot Grasping and Manipulation:

- **Sensors:** To grasp and manipulate objects effectively, robots rely on various sensors, including cameras, tactile sensors, force/torque sensors, and depth sensors, to perceive the environment and objects.

- **End Effectors:** The end effector, or robot hand, is the part of the robot responsible for grasping and manipulating objects. It can take various forms, such as grippers, suction cups, or specialized tools, depending on the application.

- **Control Algorithms:** Advanced control algorithms are crucial for coordinating the movement of the robot's joints and end effector to achieve precise grasping and manipulation. These algorithms often incorporate sensory feedback to adapt to changing conditions.

- **Planning and Perception:** Robots need planning algorithms to determine the optimal approach, grasp strategy, and path for manipulating objects. Perception systems help identify object attributes, such as shape, size, and pose, which inform the manipulation strategy.

3. Grasping Strategies:

- **Power Grasping:** This strategy involves gripping an object firmly with force, typically used for heavy or rigid objects.

- **Precision Grasping:** Precision grasping is employed for delicate and small objects, where a light touch and fine control are necessary.

- **Adaptive Grasping:** Robots can adapt their grasp based on the object's shape and properties. This may involve changing the grasp type or adjusting the grip force.

4. Challenges in Robot Grasping and Manipulation:

- **Object Variability:** Objects come in various shapes, sizes, and materials, making it challenging to develop one-size-fits-all grasping solutions.

- **Uncertainty:** Sensory noise, object slippage, and uncertain object properties require robots to handle uncertainty gracefully.

- **Real-World Interaction:** Robots must deal with complex and unstructured environments, where objects may be occluded, cluttered, or partially hidden.

- **Dexterity:** Achieving human-level dexterity in robot hands remains a significant challenge, especially for tasks like tying knots or handling fragile objects.

5. Applications of Robot Grasping and Manipulation:

- **Manufacturing:** Robots are used in assembly lines to grasp and manipulate parts for automotive, electronics, and other industries.

- **Logistics and Warehousing:** Robots play a crucial role in material handling, order fulfillment, and packaging in warehouses and distribution centers.

- **Agriculture:** Agricultural robots perform tasks such as picking fruits, harvesting crops, and pruning plants.

- **Healthcare:** Surgical robots assist in minimally invasive surgeries, while assistive robots aid individuals with disabilities.

- **Research:** Robots are used in research laboratories for tasks like sample handling, chemical analysis, and experiment automation.

6. Future Trends in Robot Grasping and Manipulation:

Advancements in machine learning, computer vision, and tactile sensing are expected to enhance robot grasping and manipulation capabilities. Robots with more sophisticated and adaptable end effectors, combined with improved perception and control algorithms, will continue to push the boundaries of what's possible in automation and robotics.

In conclusion, robot grasping and manipulation represent the epitome of robotic capabilities, enabling machines to interact with the physical world in ways that were once reserved for humans. As these skills advance, robots will become even more valuable in various industries, driving increased efficiency, precision, and versatility in a wide range of applications.

D. Mobile Robot Navigation: Navigating the Dynamic World

Mobile robot navigation is the art and science of enabling robots to autonomously move and navigate through complex and dynamic environments. It's a foundational capability that empowers robots to explore, interact, and perform tasks in diverse settings, from autonomous vehicles on city streets to warehouse robots in logistics centers. In this in-depth exploration, we will delve into the intricacies of mobile robot navigation, its methodologies, challenges, and its profound impact on various industries and technologies.

1. The Importance of Mobile Robot Navigation:

Mobile robots, which include autonomous vehicles, drones, and ground-based robots, are increasingly integrated into our daily lives and industries. Navigation is a fundamental skill that allows these robots to perform tasks such as transportation, exploration, surveillance, and delivery with minimal human intervention.

2. Key Components of Mobile Robot Navigation:

- **Sensors:** Mobile robots rely on an array of sensors, including cameras, LiDAR, GPS, IMUs (Inertial Measurement Units), ultrasonic sensors, and radar, to perceive their surroundings and gather data on obstacles, terrain, and other relevant information.

- **Localization:** Localization algorithms determine the robot's position and orientation within its environment, often using sensor data in conjunction with maps or landmarks.

- **Mapping:** Mapping techniques generate and update digital representations of the environment, allowing robots to plan paths and make informed decisions based on the map.

- **Path Planning:** Path planning algorithms compute the optimal or feasible path for the robot to follow while avoiding obstacles and adhering to constraints.

- **Control:** Control algorithms manage the robot's movements and ensure it follows the planned trajectory accurately.

3. Navigation Methods and Algorithms:

- **Simultaneous Localization and Mapping (SLAM):** SLAM is a technique that allows a robot to build a map of its environment while simultaneously estimating its own position within that map. It is crucial for navigation in unknown or partially mapped environments.

- **Reactive Navigation:** Reactive navigation relies on real-time sensor feedback to make immediate decisions, allowing robots to respond quickly to dynamic obstacles and changing environments.

- **Global and Local Path Planning:** Global planners generate high-level paths from the start to the goal, while local planners handle the details of following the path and avoiding immediate obstacles.

- **Probabilistic Approaches:** Bayesian filters, such as the Kalman filter and particle filter, are often used for probabilistic localization and navigation in uncertain environments.

4. Challenges in Mobile Robot Navigation:

- **Sensor Noise and Perception Errors:** Sensors can introduce noise, and perception errors can lead to incorrect interpretations of the environment.

- **Dynamic Environments:** Navigating in environments with moving objects or other dynamic elements requires advanced collision avoidance strategies.

- **Real-World Constraints:** Navigation must consider real-world constraints, such as vehicle dynamics, energy limitations, and safety requirements.

- **Multi-Agent Navigation:** In scenarios with multiple robots or vehicles, coordination and collision avoidance become more complex.

5. Applications of Mobile Robot Navigation:

- **Autonomous Vehicles:** Self-driving cars and drones use mobile navigation to navigate city streets, highways, and airspace.

- **Warehouse and Logistics Automation:** Robots in warehouses navigate to pick and place items, transport goods, and optimize inventory management.

- **Agricultural Robotics:** Agricultural robots navigate fields for tasks like planting, harvesting, and monitoring crops.

- **Search and Rescue:** Robots can explore disaster-stricken areas, mines, or remote locations to locate survivors or hazardous materials.

- **Surveillance and Security:** Drones and ground-based robots are used for surveillance and security patrols.

6. Future Trends in Mobile Robot Navigation:

Advancements in perception, machine learning, and AI are expected to enhance mobile robot navigation. These robots will become more capable of handling complex, unstructured environments and interacting seamlessly with humans and other robots.

In conclusion, mobile robot navigation is a cornerstone of robotics, enabling machines to navigate and interact with their environments, from the ordinary to the extraordinary. As this field continues to advance, the deployment of mobile robots in various industries and applications will become increasingly widespread, revolutionizing transportation, logistics, agriculture, and countless other domains.

E. Autonomous Control and Path Planning: Navigating the Future

Autonomous control and path planning are critical components of robotics and autonomous systems, enabling machines to operate independently in complex, dynamic environments. These capabilities empower robots and vehicles, such as autonomous cars and drones, to make intelligent decisions, navigate safely, and accomplish tasks with minimal human intervention. In this in-depth exploration, we will delve into the intricacies of autonomous control and path planning, their methodologies, challenges, and transformative impact on various industries and technologies.

1. The Essence of Autonomous Control and Path Planning:

Autonomous control and path planning are the essence of

autonomy in robots and vehicles. They allow machines to make real-time decisions and chart courses through the physical world. These technologies are vital for enabling automation in transportation, logistics, agriculture, and exploration.

2. Key Components of Autonomous Control and Path Planning:

- **Sensors:** Autonomous systems rely on sensors like LiDAR, cameras, radar, GPS, and IMUs to perceive their surroundings, gather data, and sense obstacles and terrain.

- **Localization:** Autonomous vehicles determine their precise position and orientation within their environment using techniques like GPS, SLAM (Simultaneous Localization and Mapping), or odometry.

- **Path Planning:** Path planning algorithms compute collision-free trajectories or routes for the robot or vehicle to follow, often taking into account high-level goals and environmental constraints.

- **Control:** Control systems execute the planned trajectories by adjusting the robot's actuators (e.g., motors, thrusters, or steering systems) to ensure it follows the desired path.

3. Path Planning Techniques:

- *A Search:** A* is a popular algorithm used for finding the shortest path from a start to a goal point, considering the costs associated with each possible path.

- **Dijkstra's Algorithm:** Dijkstra's algorithm is used for finding the shortest path in weighted graphs, where edge weights represent traversal costs.

- **Rapidly-exploring Random Trees (RRT):** RRT is a sampling-based algorithm that explores the configuration space efficiently to find feasible paths.

- **Model Predictive Control (MPC):** MPC is an optimization-based approach that plans control actions over a finite time horizon, considering system dynamics and constraints.

4. Challenges in Autonomous Control and Path Planning:

- **Dynamic Environments:** Navigating through environments with moving objects, pedestrians, and other vehicles introduces complexities and requires advanced collision avoidance strategies.

- **Uncertainty:** Autonomous systems must cope with sensor noise, perception errors, and uncertain environmental conditions.

- **Real-World Constraints:** Control and path planning must consider factors such as vehicle dynamics, energy limitations, and safety requirements.

- **Multi-Agent Interaction:** In scenarios with multiple autonomous vehicles or robots, coordination and collision avoidance become more intricate.

5. Applications of Autonomous Control and Path Planning:

- **Autonomous Vehicles:** Self-driving cars and drones use these technologies for navigating urban streets, highways, and airspace.

- **Delivery and Logistics:** Autonomous delivery vehicles navigate neighborhoods and warehouses for efficient package delivery.

- **Agriculture:** Autonomous tractors and drones plan paths for planting, harvesting, and monitoring crops.

- **Search and Rescue:** Autonomous robots and drones navigate disaster-stricken areas to locate survivors or hazardous materials.

- **Space Exploration:** Autonomous control and path planning are crucial for rover navigation on Mars and other celestial bodies.

6. Future Trends in Autonomous Control and Path Planning:

Advancements in AI, machine learning, and sensor technology

are poised to enhance autonomous control and path planning capabilities. Autonomous systems will become more adaptable to diverse environments, safer, and better integrated into our daily lives.

In conclusion, autonomous control and path planning are the driving forces behind the transformative potential of autonomous systems. As these technologies continue to evolve, we can anticipate a future where automation plays an increasingly integral role in transportation, logistics, agriculture, exploration, and countless other domains, making our world more efficient, connected, and safe.

F. Robotic Process Automation (RPA): Revolutionizing Workflows

Robotic Process Automation (RPA) is a transformative technology that is revolutionizing the way organizations automate routine, rule-based tasks and business processes. RPA software robots, or "bots," are designed to mimic human actions, interacting with digital systems to execute repetitive tasks with precision and efficiency. In this in-depth exploration, we will delve into the intricacies of RPA, its methodologies, benefits, challenges, and its profound impact on various industries and business processes.

1. The Essence of Robotic Process Automation (RPA):

RPA is a form of automation technology that focuses on automating repetitive and rule-based tasks within business processes. Unlike traditional automation, RPA is highly flexible, non-invasive, and can be quickly implemented without major changes to existing IT infrastructure.

2. Key Components of Robotic Process Automation (RPA):

- **RPA Bots:** RPA bots are software programs that mimic human interactions with computer systems. They can perform a wide range of tasks, including data entry, data extraction, report generation, and more.

- **Process Orchestration:** RPA often involves orchestrating multiple bots to work together to complete an end-to-end business process.

- **User Interfaces:** RPA bots interact with user interfaces in applications and systems, just like human users, to input or extract data.

- **Business Rules:** RPA relies on predefined business rules and logic to make decisions during automation processes.

3. RPA Methodologies and Technologies:

- **Screen Scraping:** Involves reading data from the screens of legacy systems or applications.

- **API Integration:** Bots can communicate with software through APIs, enabling data exchange and automation of processes that use modern applications.

- **Rule-Based Automation:** Bots follow a set of predefined rules and conditions to make decisions and complete tasks.

- **Machine Learning and AI:** Advanced RPA systems can leverage machine learning and AI to handle unstructured data and make more intelligent decisions.

4. Benefits of Robotic Process Automation (RPA):

- **Cost Efficiency:** RPA reduces operational costs by automating repetitive tasks, which reduces the need for human labor.

- **Accuracy:** Bots perform tasks with high accuracy and consistency, reducing errors and improving data quality.

- **Scalability:** RPA can be easily scaled up or down to accommodate changing workloads.

- **Enhanced Productivity:** RPA frees up human employees from mundane tasks, allowing them to focus on more value-added activities.

- **Compliance:** RPA ensures that tasks are completed according to predefined rules and regulatory requirements, reducing compliance risks.

5. Challenges in Robotic Process Automation (RPA):

- **Complex Processes:** Automating complex and non-linear processes can be challenging and may require extensive customization.

- **Data Security:** RPA systems must handle sensitive data carefully to ensure security and compliance.

- **Change Management:** Implementing RPA may require changes in organizational processes and workforce skillsets, which can face resistance.

- **Scalability Concerns:** Scaling RPA may require additional investments in infrastructure and bot management.

6. Applications of Robotic Process Automation (RPA):

- **Finance and Accounting:** RPA can automate tasks like invoice processing, accounts payable and receivable, and financial report generation.

- **Human Resources:** RPA is used for employee onboarding, payroll processing, and benefits administration.

- **Customer Service:** Bots handle customer inquiries, ticket routing, and data entry in call centers and online support systems.

- **Healthcare:** RPA assists with claims processing, patient data management, and billing in healthcare organizations.

- **Supply Chain:** RPA optimizes order processing, inventory management, and logistics tracking.

7. Future Trends in Robotic Process Automation (RPA):

As RPA continues to evolve, it will incorporate more advanced technologies like natural language processing, machine learning, and cognitive automation. This will enable bots to handle more complex tasks and unstructured data, further increasing their value to organizations.

In conclusion, Robotic Process Automation (RPA) is transforming the way organizations operate by automating routine and rule-based tasks. As this technology continues to advance, we can expect to see increased efficiency, reduced costs, and improved accuracy across various industries and business processes, making RPA a central component of modern business operations.

CHAPTER 5

Robot Learning, AI, and Artificial Intelligence in Automation: Transforming the Future of Machines

In the realm of automation and robotics, the synergy between Robot Learning, Artificial Intelligence (AI), and Artificial Intelligence in Automation is ushering in a new era of intelligent machines. These technologies are at the forefront of innovation, enabling robots to not only perform tasks but also adapt, reason, and learn from their experiences. In this introductory exploration, we will embark on a journey through the world of Robot Learning, AI, and Artificial Intelligence in Automation, uncovering their significance, applications, and the transformative potential they hold for the future of automation.

1. The Convergence of Robot Learning, AI, and Automation:

Robot Learning, AI, and Artificial Intelligence in Automation represent a convergence of technologies that empower machines to transcend their programmed capabilities. This fusion enables robots and automated systems to evolve from mere tools to intelligent collaborators in a variety of domains.

2. Key Components of Robot Learning, AI, and Artificial Intelligence in Automation:

- **Machine Learning:** Machine learning algorithms allow robots to analyze data, recognize patterns, and make decisions based on past experiences.

- **Reinforcement Learning:** Reinforcement learning enables robots to learn optimal behaviors through trial and error, similar to how humans learn.

- **Cognitive Computing:** Cognitive systems use AI to process natural language, understand context, and interact with users more intuitively.

- **Computer Vision:** Computer vision technologies enable robots to perceive and interpret visual information from their surroundings.

3. Robot Learning and AI in Automation Applications:

- **Manufacturing:** Robots equipped with AI can adapt to changing production requirements, identify defects, and optimize manufacturing processes.

- **Healthcare:** AI-powered robots assist in diagnostics, surgery, patient care, and medication management.

- **Autonomous Vehicles:** Self-driving cars utilize AI for

perception, decision-making, and navigation.

- **Service Robots:** Robots in hospitality, retail, and customer service rely on AI for natural language understanding and personalized interactions.

- **Energy and Utilities:** AI-driven automation enhances energy grid management, predictive maintenance, and resource optimization.

4. The Role of Robot Learning and AI Ethics:

As robots and AI become more integrated into daily life, ethical considerations, such as transparency, accountability, and bias mitigation, become increasingly important. Ethical frameworks and guidelines are essential to ensure the responsible development and deployment of these technologies.

5. Future Trends and Innovations:

The future of Robot Learning, AI, and Artificial Intelligence in Automation holds exciting prospects. Anticipated trends include the development of AI systems with improved generalization, enhanced robot-human collaboration, and AI models capable of continuous learning in dynamic environments.

In conclusion, Robot Learning, AI, and Artificial Intelligence in Automation are shaping a future where machines possess not just automation but also intelligence. As these technologies continue to advance, they will revolutionize industries, enhance efficiency, and redefine the relationship between humans and machines, making automation more adaptable, intelligent, and indispensable than ever before.

A. Machine Learning in Robotics: Pioneering Intelligence in Machines

Machine Learning (ML) has emerged as a transformative force in the field of robotics, equipping machines with the ability to learn from data, adapt to their environments, and perform tasks with increasing autonomy. From self-driving cars to industrial automation and healthcare robotics, ML has become an integral component in advancing the capabilities of robots. In this in-depth exploration, we will delve into the intricacies of Machine Learning in robotics, its methodologies, applications, challenges, and the groundbreaking impact it has on the world of automation.

1. The Power of Machine Learning in Robotics:

Machine Learning in robotics is the marriage of data-driven intelligence with physical automation. It allows robots to make informed decisions, recognize patterns, adapt to changes, and continuously improve their performance, making them more versatile and capable than ever before.

2. Key Components of Machine Learning in Robotics:

- **Data Acquisition:** ML models require data as input. In robotics, this data often comes from sensors, cameras, lidar, or other perception devices.

- **Training Data:** Robots are trained on datasets that contain examples of the task they are meant to perform. These datasets are labeled to provide the model with supervised learning.

- **Learning Algorithms:** ML algorithms, such as deep neural networks, decision trees, and support vector machines, process the training data and learn patterns and features relevant to the task.

- **Feedback Mechanisms:** Reinforcement learning is commonly used in robotics, where robots receive feedback (rewards or penalties) based on their actions and use this feedback to improve their behavior over time.

3. Machine Learning Applications in Robotics:

- **Autonomous Vehicles:** ML powers self-driving cars by enabling them to perceive their surroundings, make driving decisions, and learn from various road conditions.

- **Industrial Automation:** Robots in manufacturing utilize ML for quality control, predictive maintenance, and optimizing production processes.

- **Healthcare Robotics:** Surgical robots benefit from ML by enhancing precision and enabling minimally invasive procedures.

- **Agricultural Robotics:** ML helps robots in agriculture with tasks like crop monitoring, harvesting, and pest detection.

- **Service Robots:** Robots in service industries, like hospitality and retail, use ML to understand natural language, personalize experiences, and assist customers.

4. Challenges in Machine Learning for Robotics:

- **Data Quality:** High-quality data is crucial for training ML models. Noise or bias in the data can lead to incorrect learning.

- **Safety and Reliability:** Ensuring the safety of autonomous systems is a significant challenge, particularly when deploying robots in real-world environments.

- **Generalization:** ML models must generalize from their training data to perform well in diverse and unseen situations.

- **Real-Time Processing:** Many robotic applications require real-time decision-making, necessitating efficient ML algorithms.

5. Future Trends and Innovations:

The future of Machine Learning in robotics is bright, with

ongoing research and development in areas such as reinforcement learning, transfer learning, explainable AI, and human-robot collaboration. Innovations in ML hardware and software will further drive progress in this field.

In conclusion, Machine Learning in robotics represents a paradigm shift in automation, enabling machines to learn and adapt to complex and dynamic environments. As this field continues to advance, robots will become more intelligent, capable, and seamlessly integrated into our lives, revolutionizing industries and opening new horizons for automation and AI-driven robotics.

B. Reinforcement Learning for Robots: The Path to Autonomous Intelligence

Reinforcement Learning (RL) has emerged as a powerful paradigm in robotics, enabling machines to learn complex behaviors and make decisions through trial and error. This approach has proven to be particularly transformative in enabling robots to navigate dynamic and unstructured environments, perform dexterous tasks, and even master tasks that were once considered challenging for automation. In this in-depth exploration, we will delve into the intricacies of Reinforcement Learning for robots, its methodologies, applications, challenges, and the revolutionary impact it has on autonomous robotics.

1. The Essence of Reinforcement Learning for Robots:

Reinforcement Learning is a subset of machine learning that focuses on training agents to make sequences of decisions in an environment to maximize a cumulative reward. In robotics, RL provides robots with the ability to learn optimal control policies and adapt to various tasks and scenarios.

2. Key Components of Reinforcement Learning for Robots:

- **Agent:** The robot or autonomous system being trained to perform a task.

- **Environment:** The physical or virtual world in which the robot operates and interacts.

- **State:** The representation of the robot's current situation or configuration within the environment.

- **Action:** The set of possible actions or control inputs that the robot can take.

- **Reward Function:** A signal provided by the environment that indicates the immediate desirability of an action or state.

- **Policy:** The strategy or mapping from states to actions that the agent learns through RL.

3. Reinforcement Learning Methodologies:

- **Q-Learning:** A popular RL algorithm that learns the value of taking an action in a particular state and converges to an optimal policy.

- **Policy Gradient Methods:** Algorithms that learn a parameterized policy directly to maximize expected rewards.

- **Deep Reinforcement Learning (DRL):** Combining RL with deep neural networks to handle high-dimensional state and action spaces.

- **Proximal Policy Optimization (PPO):** An RL algorithm that strikes a balance between exploration and exploitation to improve training stability.

4. Reinforcement Learning Applications in Robotics:

- **Robot Navigation:** RL enables robots to navigate environments, avoiding obstacles and reaching specified goals.

- **Manipulation and Grasping:** Robots can learn to grasp objects of varying shapes and sizes with precision.

- **Autonomous Vehicles:** Self-driving cars employ RL for tasks like lane following, decision-making at intersections, and parking.

- **Game Playing:** RL algorithms have achieved superhuman performance in games like chess, Go, and video games.

- **Humanoid Robotics:** Humanoid robots utilize RL for tasks such as walking, running, and performing acrobatic maneuvers.

5. Challenges in Reinforcement Learning for Robots:

- **Sample Efficiency:** RL algorithms often require a large number of trials, which can be impractical or costly for real-world robots.

- **Safety Concerns:** Training robots through RL in real-world environments may lead to undesirable or unsafe behavior.

- **Generalization:** Ensuring that learned policies generalize to unseen situations is a significant challenge.

- **Exploration vs. Exploitation:** Balancing exploration (trying new actions) with exploitation (choosing known good actions) is crucial for effective RL.

6. Future Trends and Innovations:

The future of Reinforcement Learning for robots lies in making the training process more sample-efficient, safe, and robust. Innovations in simulation-based training, transfer learning, and curriculum learning are expected to address these challenges and further enhance the capabilities of robotic systems.

In conclusion, Reinforcement Learning is propelling robotics into an era of autonomy and adaptability. As robots continue to learn and improve their decision-making skills through RL, they will become increasingly capable of navigating complex and unpredictable environments, leading to advancements in automation, transportation, healthcare, and beyond.

C. Robot Learning and AI Ethics: Navigating the Moral Compass of Automation

Robot Learning and AI Ethics represent the ethical considerations and moral implications of integrating machine learning and artificial intelligence (AI) into robotic systems. As robots and AI technologies become more capable and autonomous, it is crucial to address the ethical challenges associated with their decision-making, behavior, and impact on society. In this in-depth exploration, we will delve into the complexities of Robot Learning and AI Ethics, examining the ethical principles, dilemmas, and strategies that guide the responsible development and deployment of intelligent robots.

1. The Ethical Imperative of Robot Learning and AI:

Robot Learning and AI Ethics stem from the recognition that robots and AI systems, like humans, can make decisions that have ethical consequences. As these technologies become increasingly autonomous, it is imperative to ensure that their actions align with human values and societal norms.

2. Key Ethical Principles in Robot Learning and AI:

- **Transparency:** Ensuring that the decision-making processes of robots and AI systems are transparent and comprehensible to humans.

- **Accountability:** Establishing mechanisms to hold both developers and autonomous systems accountable for their actions.

- **Fairness:** Mitigating bias and ensuring equitable treatment of all individuals and groups.

- **Privacy:** Respecting individuals' privacy rights and protecting sensitive data collected by robots and AI.

- **Safety:** Prioritizing the safety of humans and preventing harm in robotic interactions.

- **Beneficence:** Designing robots and AI systems to maximize overall societal benefit and minimize harm.

- **Non-Maleficence:** Ensuring that robots do not intentionally harm humans and minimize unintended negative consequences.

3. Ethical Dilemmas in Robot Learning and AI:

- **Autonomous Decision-Making:** As robots become more autonomous, they may face situations where ethical decisions need to be made, such as in autonomous vehicles during emergencies.

- **Bias and Discrimination:** Biased training data can lead to AI systems that discriminate against certain groups, reinforcing existing societal biases.

- **Privacy Violations:** Collecting and analyzing personal data for decision-making raises concerns about privacy and surveillance.

- **Human-Robot Relationships:** Ethical considerations arise when robots emulate emotions or human characteristics that can create emotional bonds.

- **Job Displacement:** Automation and robotics can lead to job displacement and economic inequality.

4. Strategies for Ensuring Ethical Robot Learning and AI:

- **Ethical Frameworks:** Developing and adhering to ethical frameworks and guidelines for designing, training, and deploying robots and AI systems.

- **Robust Testing:** Rigorous testing and validation processes to identify and mitigate ethical risks.

- **Oversight and Regulation:** Enacting regulations and oversight bodies to ensure ethical compliance in the development and use of AI and robotics.

- **Ethical by Design:** Incorporating ethical considerations into the design and development process from the outset.

- **Ethical Training Data:** Ensuring that training data is representative and free from bias to prevent discriminatory behavior.

5. The Role of AI Ethics in Human-Robot Interaction:

Ethical considerations are particularly relevant in human-robot interaction (HRI). As robots become more integrated into society, they must adhere to ethical principles to maintain trust and acceptance among humans.

6. Future Trends and Ethical Challenges:

As technology advances, new ethical challenges will emerge. These may include questions about the moral rights of robots, the implications of AI-generated content, and the ethics of human-robot relationships.

In conclusion, Robot Learning and AI Ethics are pivotal in shaping the future of automation and AI. Ensuring that robots and

AI systems act ethically and responsibly is not only a technological challenge but also a societal imperative. By addressing these ethical considerations, we can harness the transformative potential of robotics and AI while upholding human values and societal well-being.

D. Human-Robot Interaction in Automation: The Convergence of Technology and Humanity

Human-Robot Interaction (HRI) in automation represents the dynamic interplay between humans and robots as they collaborate, communicate, and coexist in various domains. As automation technologies advance, the importance of effective HRI becomes increasingly critical. It encompasses not only the physical interaction between humans and robots but also the design of interfaces, communication modalities, and social aspects of these interactions. In this in-depth exploration, we will delve into the intricacies of HRI in automation, its methodologies, applications, challenges, and the transformative impact it has on the evolving relationship between humans and machines.

1. The Significance of Human-Robot Interaction in Automation:

HRI in automation plays a pivotal role in making technology accessible, intuitive, and beneficial to users. It enables robots and automated systems to integrate seamlessly into various industries and settings, enhancing productivity and improving the quality of life.

2. Key Components of Human-Robot Interaction in Automation:

- **Physical Interaction:** The physical contact and collaboration between humans and robots, such as cooperative assembly tasks in manufacturing.

- **Interface Design:** Designing intuitive and user-friendly interfaces that enable humans to interact with robots through graphical displays, touchscreens, or voice commands.

- **Communication Modalities:** Determining how robots communicate with humans, including speech, gestures, facial expressions, and text-based interactions.

- **Social Interaction:** Understanding and facilitating the social aspects of HRI, including trust, empathy, and ethical considerations.

3. Methodologies and Approaches in HRI:

- **Wizard of Oz Studies:** Researchers sometimes employ human "wizards" behind the scenes to control the robot's behavior and simulate HRI scenarios for evaluation.

- **User-Centered Design:** Involving end-users in the design and evaluation process to create systems that meet their needs and preferences.

- **Natural Language Processing:** Advancements in NLP enable robots to understand and respond to spoken language, facilitating more natural and intuitive interactions.

- **Human-Robot Teaming:** Collaborative robots, or cobots, are designed to work alongside humans in shared workspaces, requiring sophisticated HRI capabilities.

4. Applications of Human-Robot Interaction in Automation:

- **Manufacturing:** Robots collaborate with human workers in assembly, welding, and quality control, improving efficiency and precision.

- **Healthcare:** Surgical robots assist surgeons during minimally invasive procedures, and robotic exoskeletons aid in rehabilitation.

- **Customer Service:** Chatbots and virtual assistants provide customer support and enhance user experiences in various industries.

- **Autonomous Vehicles:** HRI is crucial in autonomous vehicles, ensuring smooth interactions between passengers and the vehicle's AI system.

- **Education and Entertainment:** Social robots are employed in educational settings and entertainment industries, providing companionship and assistance.

5. Challenges in Human-Robot Interaction in Automation:

- **Trust and Acceptance:** Building trust between humans and robots is essential for acceptance and effective collaboration.

- **Safety:** Ensuring the safety of humans when interacting with robots, especially in shared workspaces.

- **Ethical Considerations:** Addressing ethical questions, such as robot rights, privacy, and the potential for misuse.

- **Cultural Differences:** HRI must account for cultural variations in communication and social norms.

- **Usability and Accessibility:** Designing interfaces that are accessible to all users, including those with disabilities.

6. Future Trends and Innovations:

The future of HRI in automation holds exciting possibilities, including improved natural language understanding, emotion recognition, and the development of social robots that can understand and respond to human emotions.

In conclusion, Human-Robot Interaction in automation is not only shaping the way we work and live but also redefining the relationship between humans and machines. By addressing the challenges and opportunities in HRI, we can harness the full potential of automation technologies, creating a future where robots and humans collaborate harmoniously for the betterment of society.

E. AI in Automation (Artificial Intelligence and Automation): Transforming Industries and Workflows

AI in Automation represents the convergence of Artificial Intelligence (AI) technologies with automation processes, reshaping industries, and revolutionizing workflows across various domains. By integrating AI capabilities such as machine learning, computer vision, natural language processing, and decision-making algorithms, automation systems become more intelligent, adaptive, and efficient. In this in-depth exploration, we will delve into the intricacies of AI in Automation, its methodologies, applications, challenges, and the transformative impact it has on modern business processes and industries.

1. The Synergy of AI and Automation:

AI and Automation together empower machines to perform tasks that traditionally required human intelligence, decision-making, and adaptability. By infusing automation processes with AI, organizations can streamline operations, reduce costs, enhance productivity, and unlock new opportunities.

2. Key Components of AI in Automation:

- **Machine Learning:** AI systems can learn from data to make predictions, classify information, and optimize processes.

- **Computer Vision:** AI-powered visual perception enables automation systems to analyze and interpret images and videos, facilitating tasks like object recognition and quality control.

- **Natural Language Processing (NLP):** AI algorithms process and understand human language, enabling automation systems to interact with users through speech or text.

- **Cognitive Computing:** Combining AI with human-like reasoning and problem-solving capabilities to automate complex tasks.

- **Robotic Process Automation (RPA):** Software robots execute rule-based tasks, automating repetitive processes without human intervention.

3. Methodologies and Approaches in AI in Automation:

- **Supervised Learning:** AI systems are trained on labeled data to make predictions or classifications, often used in quality control and decision support.

- **Unsupervised Learning:** AI algorithms identify patterns and relationships in data without explicit labels, useful for data clustering and anomaly detection.

- **Reinforcement Learning:** AI systems learn optimal actions through trial and error, employed in optimizing processes and autonomous decision-making.

- **Deep Learning:** Neural networks with multiple layers are used in complex tasks like image recognition, language translation, and speech synthesis.

4. Applications of AI in Automation:

- **Manufacturing:** AI-driven automation improves quality control, predictive maintenance, and production optimization.

- **Healthcare:** AI assists in medical diagnosis, drug discovery, and patient care, increasing efficiency and accuracy.

- **Finance:** AI automates fraud detection, risk assessment, and algorithmic trading in financial institutions.

- **Customer Service:** Chatbots and virtual assistants provide 24/7 support, improving customer experiences.

- **Supply Chain:** AI optimizes inventory management, demand forecasting, and logistics.

5. Challenges in AI in Automation:

- **Data Quality:** AI systems rely on high-quality data, and poor data quality can lead to inaccurate results.

- **Ethical Concerns:** Ensuring AI systems make ethical decisions and avoid biases is crucial.

- **Integration Complexity:** Integrating AI into existing automation systems can be complex and costly.

- **Cybersecurity:** Protecting AI-driven automation from cyber threats and attacks is a growing concern.

6. Future Trends and Innovations:

The future of AI in Automation holds exciting developments, including improved AI interpretability, autonomous systems with AI reasoning capabilities, and increased use of AI in edge computing for real-time decision-making.

In conclusion, AI in Automation represents a paradigm shift in how organizations operate and leverage technology. As AI continues to advance, the integration of intelligent automation into various industries will enhance efficiency, reduce errors, and enable more sophisticated decision-making, ultimately reshaping the future of work and industry.

CHAPTER 6

Robot Vision, Perception, and Cognitive Automation: The Senses and Mind of Intelligent Machines

In the realm of automation and robotics, the fusion of advanced visual perception, cognitive capabilities, and automation technologies has given rise to a new era of intelligent machines. Robot Vision, Perception, and Cognitive Automation represent the marriage of sensory perception and cognitive reasoning, enabling robots to perceive, understand, and interact with their environment in ways that mimic human sensory and cognitive functions. In this introductory exploration, we embark on a journey through the fascinating world of Robot Vision, Perception, and Cognitive Automation, uncovering their significance, applications, and the transformative potential they hold for automation and robotics.

A. 3D Perception and Reconstruction: Enabling Robots to See and Understand the World in Three Dimensions

3D perception and reconstruction are integral components of Robot Vision, Perception, and Cognitive Automation, providing machines with the ability to perceive and comprehend their environment in three dimensions. This capability allows robots to navigate complex spaces, interact with objects, and make informed decisions based on a richer understanding of the world around them. In this in-depth exploration, we will delve into the intricacies of 3D perception and reconstruction, examining their methodologies, applications, challenges, and the transformative impact they have on the perception and cognition of intelligent robots.

1. The Significance of 3D Perception and Reconstruction:

3D perception and reconstruction are fundamental for robots to operate effectively in real-world environments. They go beyond traditional 2D vision, enabling machines to perceive depth, shape, and spatial relationships, which are essential for tasks like object manipulation, navigation, and scene understanding.

2. Key Components of 3D Perception and Reconstruction:

- **Sensors:** Hardware devices such as depth cameras (e.g., LiDAR), stereo cameras, structured light sensors, and time-of-flight cameras capture 3D information from the environment.

- **Point Clouds:** 3D data is often represented as point clouds, where each point corresponds to a 3D coordinate in space.

- **Computer Vision Algorithms:** Advanced algorithms process the 3D data to extract features, detect objects, and reconstruct the environment.

- **Simultaneous Localization and Mapping (SLAM):** SLAM algorithms combine 3D perception with robot localization, enabling the creation of maps and navigation in unknown environments.

3. Methodologies and Approaches in 3D Perception and Reconstruction:

- **Stereo Vision:** Depth information is extracted by comparing images from two cameras with known baseline distances.

- **Structured Light Scanning:** Patterns of light are projected onto objects, and their deformation is used to calculate depth.

- **LiDAR Scanning:** LiDAR sensors emit laser beams and measure the time it takes for the beams to bounce back, creating a 3D point cloud.

- **Depth from Motion:** Utilizes the motion of the robot or camera to estimate depth from parallax or motion parallax.

4. Applications of 3D Perception and Reconstruction:

- **Robot Navigation:** Robots use 3D perception to avoid obstacles and plan collision-free paths.

- **Object Manipulation:** Precise 3D perception is crucial for robots to grasp and manipulate objects with varying shapes and sizes.

- **Augmented Reality (AR):** AR applications overlay digital information onto the real world, relying on accurate 3D scene reconstruction.

- **Autonomous Vehicles:** Self-driving cars use 3D perception to perceive the road, detect obstacles, and navigate safely.

- **Medical Imaging:** 3D imaging technologies enhance medical diagnostics, surgical planning, and treatment.

5. Challenges in 3D Perception and Reconstruction:

- **Sensor Limitations:** Sensor noise, occlusions, and environmental conditions can affect the quality of 3D data.

- **Computational Complexity:** Processing and interpreting large 3D point clouds in real-time can be computationally intensive.

- **Integration:** Integrating 3D perception into robotic systems and ensuring accurate calibration is challenging.

- **Data Fusion:** Combining 3D data from multiple sensors and modalities for a comprehensive understanding of the environment.

6. Future Trends and Innovations:

The future of 3D perception and reconstruction lies in enhancing the robustness and efficiency of algorithms, improving sensor technologies, and enabling robots to learn from 3D data for better decision-making and adaptability.

In conclusion, 3D perception and reconstruction are foundational technologies that empower robots to interact with the world in a manner more akin to human perception. As these technologies continue to advance, robots will become more capable, versatile, and integrated into various industries, ushering in a new era of automation and cognitive automation.

B. Object Recognition and Tracking: The Eyes and Memory of Intelligent Machines

Object recognition and tracking are pivotal components of Robot Vision, Perception, and Cognitive Automation, endowing machines with the ability to identify and monitor objects within their visual field. These capabilities are essential for various applications, from autonomous navigation and surveillance to augmented reality and robotics. In this in-depth exploration, we will delve into the intricacies of object recognition and tracking, examining their methodologies, applications, challenges, and the transformative impact they have on the perception and cognition of intelligent machines.

1. The Significance of Object Recognition and Tracking:

Object recognition and tracking serve as the visual senses and memory of intelligent machines. By enabling robots and automated systems to identify and monitor objects, they can make informed decisions, respond to dynamic environments, and interact effectively with their surroundings.

2. Key Components of Object Recognition and Tracking:

- **Object Detection:** The process of identifying and locating objects of interest within an image or video stream.

- **Object Classification:** Assigning labels or categories to recognized objects, such as identifying whether an object is a car, a person, or a specific type of fruit.

- **Object Tracking:** Continuously monitoring the position, movement, and identity of objects over time.

- **Feature Extraction:** Extracting distinctive visual features from objects, such as edges, corners, or texture patterns, for recognition and tracking.

3. Methodologies and Approaches in Object Recognition and Tracking:

- **Deep Learning:** Convolutional Neural Networks (CNNs) have revolutionized object recognition, enabling the development of highly accurate and robust models.

- **Object Tracking Algorithms:** Various tracking algorithms, such as Kalman filters, Particle filters, and Mean-Shift tracking, are used to follow objects through successive frames of video.

- **Feature-Based Recognition:** Identifying objects based on their unique visual features or keypoints.

- **Template Matching:** Comparing regions of an image with predefined templates to detect objects.

4. Applications of Object Recognition and Tracking:

- **Autonomous Vehicles:** Object recognition helps self-driving cars detect and classify other vehicles, pedestrians, and obstacles.

- **Surveillance and Security:** Tracking individuals and objects in security camera footage for threat detection and monitoring.

- **Augmented Reality (AR):** Overlaying digital information onto the real world, enhancing gaming, navigation, and information retrieval.

- **Robotics:** Robots use object recognition and tracking for tasks such as grasping objects, following humans, and navigating cluttered environments.

- **Healthcare:** Object recognition assists in medical imaging, including the identification of tumors and anomalies.

5. Challenges in Object Recognition and Tracking:

- **Variability:** Objects can vary in appearance due to changes in lighting, viewpoint, occlusions, and deformations.

- **Real-Time Processing:** Achieving real-time object recognition and tracking can be computationally demanding.

- **Object Occlusion:** When objects are partially or fully hidden from view, maintaining tracking can be challenging.

- **Scale and Complexity:** Handling a large number of objects and complex scenes requires robust algorithms.

6. Future Trends and Innovations:

The future of object recognition and tracking lies in improving accuracy, robustness, and adaptability through advanced deep learning models, multi-modal sensing, and integration with other perception and cognition capabilities.

In conclusion, object recognition and tracking are fundamental technologies that empower machines to understand and interact with the visual world. As these technologies continue to advance, robots, autonomous systems, and augmented reality applications will become more capable, enhancing safety, efficiency, and user experiences in a wide range of domains.

C. Visual SLAM (Simultaneous Localization and Mapping): Navigating the Uncharted Territory of Autonomy

Visual SLAM, or Simultaneous Localization and Mapping, represents a cutting-edge technology that allows robots, autonomous vehicles, and augmented reality systems to navigate and interact with the physical world by simultaneously building maps of their surroundings and determining their own positions within those maps. Visual SLAM combines computer vision, sensor fusion, and probabilistic modeling to enable machines to operate autonomously in dynamic and unstructured environments. In this in-depth exploration, we will delve into the intricacies of Visual SLAM, examining its methodologies, applications, challenges, and the transformative impact it has on the field of autonomous systems and robotics.

1. The Significance of Visual SLAM:

Visual SLAM is a critical technology for enabling autonomous machines to understand and interact with their surroundings. It finds applications in robotics, autonomous navigation, augmented reality, and more, providing the foundation for precise localization and mapping capabilities.

2. Key Components of Visual SLAM:

- **Visual Sensors:** Cameras are the primary sensors used in Visual SLAM, capturing images or video sequences of the environment.

- **Feature Extraction:** Algorithms detect and track distinctive visual features, such as corners, edges, and keypoints, across consecutive frames.

- **Map Representation:** Visual SLAM systems maintain a map of the environment, which can include 3D point clouds, keyframes, or occupancy grids.

- **Pose Estimation:** Estimating the position and orientation of the camera or sensor relative to the mapped environment is crucial for localization.

- **Loop Closure Detection:** Recognizing previously visited locations helps correct accumulated errors and refine the map.

3. Methodologies and Approaches in Visual SLAM:

- **Feature-Based SLAM:** This approach relies on detecting and matching visual features across frames to estimate camera motion and build a map.

- **Direct SLAM:** Direct methods optimize the alignment of pixel intensities between consecutive frames, often resulting in dense 3D maps.

- **ORB-SLAM, LSD-SLAM, and SVO:** These are examples of popular Visual SLAM systems that implement various algorithms and optimizations.

- **RGB-D SLAM:** Combining color and depth information from sensors like the Microsoft Kinect to improve mapping and localization accuracy.

4. Applications of Visual SLAM:

- **Autonomous Vehicles:** Visual SLAM enables self-driving cars to navigate urban environments and highways.

- **Robotics:** Robots use Visual SLAM for tasks like autonomous exploration, mapping, and object manipulation.

- **Augmented Reality (AR):** AR applications overlay digital content onto the real world, requiring accurate pose estimation.

- **Drones:** Visual SLAM allows drones to fly autonomously, avoid obstacles, and create 3D maps.

- **Virtual Reality (VR):** VR headsets can incorporate Visual SLAM to track the user's movement and provide a more immersive experience.

5. Challenges in Visual SLAM:

- **Scale and Complexity:** Handling large-scale environments with varying lighting conditions and dynamic objects is challenging.

- **Real-Time Processing:** Achieving real-time performance on resource-constrained devices can be demanding.

- **Loop Closures and Drift:** Maintaining accurate maps and preventing accumulated errors over time.

- **Robustness to Occlusions:** Handling situations where parts of the environment are temporarily or permanently hidden.

6. Future Trends and Innovations:

The future of Visual SLAM lies in improving robustness, scalability, and efficiency through advanced deep learning techniques, sensor fusion with LiDAR and other sensors, and its integration into a wide range of autonomous systems.

In conclusion, Visual SLAM is a cornerstone technology for enabling machines to understand and interact with their surroundings autonomously. As it continues to advance, Visual SLAM will play a pivotal role in shaping the future of autonomous vehicles, robotics, augmented reality, and other fields, unlocking new possibilities for automation and human-machine interaction.

D. Object Manipulation with Vision: Bridging the Gap Between Perception and Action

Object manipulation with vision represents a sophisticated integration of robotic perception and control, allowing machines to not only recognize objects in their environment but also interact with them in a purposeful and dexterous manner. This capability is essential for a wide range of applications, from industrial automation to service robots and healthcare. In this in-depth exploration, we will delve into the intricacies of object manipulation with vision, examining its methodologies, applications, challenges, and the transformative impact it has on robotics and automation.

1. The Significance of Object Manipulation with Vision:

Object manipulation with vision is the bridge between a robot's ability to perceive its surroundings and its capability to physically interact with objects. It enables robots to perform tasks such as picking up objects, assembling components, and even assisting with delicate surgical procedures.

2. Key Components of Object Manipulation with Vision:

- **Vision Systems:** Cameras and depth sensors capture visual data about the environment and objects.

- **Object Recognition:** Algorithms identify and locate objects within the robot's field of view.

- **Gripper or Manipulator:** The robotic arm or end-effector used for physically interacting with objects.

- **Control Algorithms:** These algorithms determine how the robot should grasp, lift, or manipulate objects based on visual data.

3. Methodologies and Approaches in Object Manipulation with Vision:

- **Grasping Algorithms:** These determine the optimal way to grasp an object, considering factors like shape, size, and weight.

- **Visual Servoing:** A control technique that uses visual feedback to adjust the robot's motion during manipulation tasks.

- **Reinforcement Learning:** Training robots to manipulate objects through trial and error, learning from both success and failure.

- **Simulated Environments:** Using simulated environments for training and testing manipulation algorithms before deploying them in the real world.

4. Applications of Object Manipulation with Vision:

- **Manufacturing:** Robots equipped with vision systems

manipulate objects in assembly lines, enhancing production efficiency.

- **Logistics and Warehousing:** Autonomous robots pick and place items in warehouses and logistics centers.

- **Healthcare:** Surgical robots assist surgeons in delicate procedures, improving precision and reducing invasiveness.

- **Agriculture:** Robots equipped with vision systems can harvest crops and perform other tasks on farms.

- **Service Robots:** Robots in hospitality and retail settings handle objects like dishes, packages, and merchandise.

5. Challenges in Object Manipulation with Vision:

- **Object Variability:** Objects can vary in shape, size, and texture, making recognition and manipulation challenging.

- **Real-World Clutter:** Dealing with cluttered environments and occlusions during manipulation tasks.

- **Robotic Grasping:** Ensuring the robot's grasp is stable and secure, even for irregularly shaped objects.

- **Perception-Action Integration:** Coordinating perception and action in real-time to adapt to changing situations.

6. Future Trends and Innovations:

The future of object manipulation with vision lies in improving robustness, adaptability, and versatility through advancements in deep learning, multimodal sensing, and haptic feedback integration.

In conclusion, object manipulation with vision represents a critical advancement in robotics and automation. As these technologies continue to evolve, robots will become more capable and versatile in handling a wide range of tasks, contributing to increased productivity and safety across various industries.

E. Augmented Reality (AR) and Robotics in Automation: Transforming Industries and Workflows

Augmented Reality (AR) and Robotics, when combined, offer a powerful synergy that has the potential to revolutionize industries and workflows. AR overlays digital information onto the physical world, enhancing human perception and interaction, while robotics automates tasks with precision and efficiency. Together, they create a paradigm where robots are not just tools but intelligent entities that interact with humans in a dynamic and augmented environment. In this in-depth exploration, we will delve into the intricacies of AR and Robotics in Automation, examining their methodologies, applications, challenges, and the transformative impact they have on modern industry and work processes.

1. The Significance of AR and Robotics in Automation:

AR and Robotics in Automation represent a convergence of technologies that improves human-machine interaction, efficiency, safety, and the ability to handle complex tasks. This fusion is particularly valuable in industrial settings, healthcare, education, and entertainment.

2. Key Components of AR and Robotics in Automation:

- **AR Devices:** Hardware like AR glasses, smartphones, and tablets equipped with cameras and displays are used to visualize augmented content.

- **Robotic Systems:** Robots with sensors, actuators, and computing capabilities are responsible for task execution and interaction.

- **Computer Vision:** Vision algorithms enable robots to perceive and interact with the real world, recognizing objects and tracking their movements.

- **Interaction Interfaces:** User interfaces that allow users to interact with AR content and control robots, often through gestures, voice, or touch.

3. Methodologies and Approaches in AR and Robotics in Automation:

- **Marker-Based AR:** AR content is anchored to physical markers or objects, allowing precise alignment of digital information.

- **Markerless AR:** AR content is placed in the real world based on environmental features and object recognition.

- **Simultaneous Localization and Mapping (SLAM):** Robotics use SLAM techniques to create maps of the environment and navigate with precision.

- **Object Tracking:** Robots track objects in real-time, allowing them to interact with and manipulate them more accurately.

4. Applications of AR and Robotics in Automation:

- **Manufacturing:** AR guides assembly line workers with step-by-step instructions, while robots handle repetitive tasks with precision.

- **Healthcare:** Surgeons use AR overlays during procedures, and robotic systems assist in surgeries and rehabilitation.

- **Education:** AR enhances learning experiences by providing interactive, 3D educational content, and robots facilitate hands-on learning.

- **Entertainment:** AR and robotics create immersive gaming experiences and interactive attractions in theme parks.

- **Maintenance and Repair:** Field technicians use AR to access repair instructions and robotics for remote assistance.

5. Challenges in AR and Robotics in Automation:

- **Integration Complexity:** Integrating AR systems with robotic platforms can be technically challenging and costly.

- **Data Security:** Safeguarding sensitive data transmitted through AR and robotic systems is crucial.

- **User Acceptance:** Ensuring that users are comfortable with AR and robotic interactions, addressing potential resistance or skepticism.

- **Environmental Variability:** AR systems may struggle in varying lighting and environmental conditions, affecting performance.

6. Future Trends and Innovations:

The future of AR and Robotics in Automation lies in advancing the sophistication of AR interfaces, improving robotic dexterity, and enabling greater autonomy through AI and machine learning.

In conclusion, the fusion of AR and Robotics in Automation has the potential to redefine how humans interact with machines and how industries operate. As these technologies continue to mature and find wider adoption, we can anticipate increased productivity, safety, and innovation across numerous sectors, shaping the future of automation and human-machine collaboration.

CHAPTER 7

Robot Control, Programming, and Business Process Automation (BPA): Orchestrating Efficiency and Precision in the Digital Age

Robot Control, Programming, and Business Process Automation (BPA) represent the command center of modern automation, where intelligent machines are directed to perform tasks with precision and speed. This convergence of robotics, programming, and automation technologies is not only reshaping industrial landscapes but also revolutionizing how businesses streamline their operations. In this introductory exploration, we embark on a journey into the world of Robot Control, Programming, and Business Process Automation, uncovering their significance, applications, and the transformative potential they hold for industries and enterprises in the digital age.

A. Robot Programming Languages (ROS, Python, C++): The Code Behind Autonomous Machines

Robot programming languages are the lifeblood of intelligent machines, empowering them to execute tasks, make decisions, and interact with the world. In the realm of robotics, several programming languages have gained prominence, each with its strengths and applications. This in-depth exploration focuses on three key languages: ROS (Robot Operating System), Python, and C++. We will delve into their methodologies, applications, and the role they play in shaping the capabilities and autonomy of robots.

1. ROS (Robot Operating System):

- **Significance:** ROS is not just an operating system but a middleware framework that simplifies the development of robot software. It provides libraries and tools for tasks like hardware abstraction, device drivers, communication between processes, and package management.

- **Methodology:** ROS follows a modular architecture, where software is organized into nodes that communicate through a publish-subscribe mechanism. It encourages code reuse and collaboration among the robotics community.

- **Applications:** ROS is widely used in research, industrial robotics, and autonomous systems development. It simplifies the integration of sensors, actuators, and complex algorithms, making it a go-to choice for robot developers.

2. Python:

- **Significance:** Python is a versatile, high-level programming language known for its readability and ease of use. It has a strong community and a vast ecosystem of libraries and frameworks that make it suitable for rapid prototyping and development.

- **Methodology:** Python offers simplicity and expressiveness, making it an excellent choice for scripting tasks, control logic, and high-level decision-making in robotics. It can be used in conjunction with other languages like C++ for performance-critical components.

- **Applications:** Python is employed in various robotics applications, including autonomous navigation, computer vision, and machine learning. Its simplicity and versatility make it accessible to both beginners and experts in robotics.

3. C++:

- **Significance:** C++ is a low-level, high-performance programming language known for its efficiency and control over hardware resources. It is often used in situations where real-time performance and low-level control are essential.

- **Methodology:** C++ is favored for developing critical components of robotic systems, such as motion control, sensor drivers, and algorithms that require fine-grained memory management. It excels in applications where speed and efficiency are paramount.

- **Applications:** C++ is commonly used in robotics for tasks like sensor data processing, real-time control, and embedded systems. It underpins many robotics libraries and frameworks for performance-critical code.

4. Choosing the Right Language:

- **Task Requirements:** The choice of programming language depends on the specific requirements of the robotic task. Python may be preferred for high-level control and scripting, while C++ may be necessary for real-time processing and resource-intensive operations.

- **Integration:** Many robotics projects involve a combination of languages. ROS, for example, allows nodes to be written in different languages and communicate seamlessly, enabling developers to leverage the strengths of each language.

- **Development Team:** The expertise of the development team plays a role in language selection. A team skilled in Python may opt for that language, while a team with a background in embedded systems may lean toward C++.

In conclusion, the choice of programming language in robotics is influenced by a combination of factors, including the nature of the task, performance requirements, and the expertise of the development team. The synergy between languages like ROS, Python, and C++ enables robots to exhibit autonomy, intelligence, and adaptability in a wide range of applications, shaping the future of automation and robotics.

B. Real-time Control in Robotics and Automation: Navigating the Speed of Precision

Real-time control is the heartbeat of robotics and automation, ensuring that machines respond swiftly and accurately to changing conditions and inputs. It's the difference between a robot deftly assembling complex components and a drone autonomously navigating through a dynamic environment. In this in-depth exploration, we will delve into the intricacies of real-time control in robotics and automation, examining its methodologies, applications, challenges, and the transformative impact it has on achieving precision and efficiency in various industries.

1. Significance of Real-time Control:

- **Precision:** Real-time control allows robots and automated systems to perform tasks with millisecond-level precision, critical for applications like manufacturing, surgery, and autonomous vehicles.

- **Safety:** In scenarios where human safety is paramount, real-time control ensures rapid responses to unforeseen events, mitigating potential risks.

- **Efficiency:** Many industrial processes and automated systems require real-time optimization to achieve maximum efficiency and productivity.

2. Key Components of Real-time Control:

- **Sensors:** Sensors, such as cameras, LiDAR, and encoders, continuously feed data to the control system, allowing it to react to changes in the environment.

- **Control Algorithms:** These algorithms process sensor data and generate control signals, adjusting the machine's actions in real-time.

- **Actuators:** Motors, servos, and other actuators execute the control commands, translating digital instructions into physical movements.

- **Feedback Loops:** Closed-loop control systems use feedback from sensors to adjust control signals continuously, ensuring that the desired state is achieved and maintained.

3. Methodologies and Approaches in Real-time Control:

- **PID Control:** Proportional-Integral-Derivative control is a common approach used in real-time control systems, adjusting control signals based on the error between the desired and actual state.

- **Model Predictive Control (MPC):** MPC uses mathematical models of the system to predict future states and optimize control signals accordingly, often used in autonomous vehicles and industrial processes.

- **Fuzzy Logic Control:** Fuzzy logic allows for the incorporation of linguistic variables and human-like reasoning into control systems, useful in situations with complex, uncertain inputs.

4. Applications of Real-time Control:

- **Manufacturing:** Real-time control ensures precision and synchronization in assembly lines and machining processes.

- **Autonomous Vehicles:** Self-driving cars and drones rely on real-time control to navigate through dynamic environments and avoid obstacles.

- **Medical Robotics:** Surgical robots and medical devices use real-time control for precise, minimally invasive procedures.

- **Aerospace:** Aircraft and spacecraft require real-time control for flight stability, navigation, and landing.

- **Process Control:** In industries like chemical manufacturing, real-time control optimizes processes and ensures safety.

5. Challenges in Real-time Control:

- **Latency:** Minimizing communication and processing latency is crucial for real-time systems, as delays can lead to inaccuracies or safety risks.

- **Synchronization:** Coordinating the actions of multiple actuators and sensors in a real-time system can be challenging.

- **Hardware Constraints:** Hardware limitations, such as processor speed and sensor accuracy, can affect the real-time performance of control systems.

6. Future Trends and Innovations:

The future of real-time control lies in leveraging advances in machine learning and artificial intelligence to create adaptive, self-learning control systems that can handle complex and dynamic environments with even greater precision and efficiency.

In conclusion, real-time control is the driving force behind the precision and efficiency of robotics and automation. As technology continues to evolve, real-time control systems will become more sophisticated and adaptable, enabling machines to perform increasingly complex tasks and further transforming industries and daily life.

C. Behavior-Based Robotics: Building Intelligence Through Simplicity and Adaptability

Behavior-based robotics is a paradigm that emphasizes the development of intelligent behavior in robots through the combination of simple, reactive modules or behaviors. Unlike traditional AI approaches, behavior-based robotics forgoes complex internal representations and instead focuses on the direct coupling of sensors to actions. In this in-depth exploration, we will delve into the intricacies of behavior-based robotics, examining its methodologies, applications, challenges, and the transformative impact it has on creating adaptable and intelligent robotic systems.

1. Significance of Behavior-Based Robotics:

- **Simplicity:** Behavior-based robotics seeks to simplify the design of intelligent systems by breaking down complex behaviors into manageable, modular components.

- **Adaptability:** This approach allows robots to adapt to changing environments and tasks more effectively, as behaviors can be easily combined or modified.

- **Robustness:** Behavior-based systems are often robust in unstructured or dynamic settings, as they react directly to sensory input without relying on elaborate planning or reasoning.

2. Key Components of Behavior-Based Robotics:

- **Behaviors:** Behaviors are the fundamental building blocks of behavior-based robots. Each behavior is responsible for a specific task or reaction to sensor input.

- **Sensors:** Sensors provide information about the robot's environment, including data on proximity, light, sound, or touch.

- **Actuators:** Actuators are the mechanisms responsible for executing the robot's actions, such as motors, wheels, or grippers.

- **Arbitration:** An arbitration mechanism determines which behavior is currently active or should take precedence based on the sensor input and the robot's goals.

3. Methodologies and Approaches in Behavior-Based Robotics:

- **Subsumption Architecture:** Developed by Rodney Brooks, the subsumption architecture organizes behaviors into layers, with higher-level behaviors subsuming lower-level ones. This allows for emergent and adaptive behavior.

- **Reactive Control:** Behaviors in this approach react directly to sensory input without the need for internal state representation or planning.

- **Hybrid Systems:** Combining behavior-based approaches with traditional AI methods like planning or learning to create more versatile robotic systems.

4. Applications of Behavior-Based Robotics:

- **Search and Rescue:** Behavior-based robots can navigate disaster-stricken areas, detect survivors, and coordinate rescue efforts.

- **Agriculture:** Robots equipped with behavior-based systems can autonomously navigate fields, detect and treat pests, and perform precision farming tasks.

- **Exploration:** Planetary rovers and autonomous drones often rely on behavior-based control for adapting to unknown terrains and dynamic conditions.

- **Education and Research:** Behavior-based robotics is commonly used in educational robotics platforms to teach principles of robotics and programming.

5. Challenges in Behavior-Based Robotics:

- **Integration:** Combining multiple behaviors and ensuring they work harmoniously can be challenging.

- **Fine-Tuning:** Designing and fine-tuning behaviors to achieve desired outcomes can be time-consuming.

- **Complex Behaviors:** Implementing complex behaviors may require substantial effort in breaking them down into simpler components.

6. Future Trends and Innovations:

The future of behavior-based robotics may involve the integration of learning and adaptation mechanisms to allow robots to improve their behaviors over time.

In conclusion, behavior-based robotics represents a pragmatic approach to building intelligent robots by focusing on simplicity, adaptability, and direct sensor-action coupling. As robotics continues to evolve, this paradigm will play a pivotal role in creating robots that can navigate, interact, and adapt in a wide range of real-world environments and tasks.

D. Robot Simulation and Testing: The Crucible of Autonomous Machines

Robot simulation and testing are critical phases in the development and deployment of autonomous machines. They serve as virtual proving grounds where robots are rigorously tested, fine-tuned, and validated before they interact with the physical world. In this in-depth exploration, we will delve into the intricacies of robot simulation and testing, examining their methodologies, applications, challenges, and the transformative impact they have on ensuring the safety and reliability of robotic systems.

1. Significance of Robot Simulation and Testing:

- **Safety Assurance:** Simulation allows developers to identify and rectify potential safety hazards and malfunctions without risking damage to physical robots or harm to humans.

- **Cost-Efficiency:** Testing in a virtual environment is more cost-effective than building and maintaining physical prototypes.

- **Iterative Development:** Simulations enable rapid iterations and refinements of robot designs and algorithms, accelerating the development process.

2. Key Components of Robot Simulation and Testing:

- **Simulation Environment:** A digital replica of the robot's operating environment, including objects, terrain, and obstacles.

- **Physics Engine:** Simulates the physical interactions between the robot and its environment, ensuring realism in movements and collisions.

- **Sensor Models:** Accurate models of sensors such as cameras, LiDAR, and ultrasonic sensors to replicate real-world data acquisition.

- **Control Software:** The same control software used on physical robots is typically run in the simulated environment.

3. Methodologies and Approaches in Robot Simulation and Testing:

- **Hardware-in-the-Loop (HIL):** Integrates real hardware components, like sensors and actuators, with the simulation to test specific aspects of the robot's performance.

- **Software-in-the-Loop (SIL):** Focuses on testing the robot's control software within the simulation environment, often used for algorithm validation.

- **Closed-Loop Testing:** Simulates real-world scenarios where the robot interacts with the environment and other virtual entities, assessing its decision-making and execution.

- **Benchmarking:** Comparing the performance of different algorithms or robot configurations within the same simulation environment to determine the most effective solutions.

4. Applications of Robot Simulation and Testing:

- **Autonomous Vehicles:** Simulations enable the testing of self-driving car algorithms in various driving scenarios without the risks associated with real-world testing.

- **Manufacturing:** Robots used in manufacturing can be optimized for efficiency and safety in a simulated factory environment.

- **Space Exploration:** Space agencies use simulations to test the autonomy and navigation of robotic spacecraft and rovers.

- **Robotics Research:** Simulations are indispensable in robotics research to evaluate novel algorithms and approaches.

5. Challenges in Robot Simulation and Testing:

- **Fidelity:** Achieving high-fidelity simulations that accurately replicate real-world conditions can be challenging.

- **Sensor Models:** Creating accurate sensor models that mimic the noise and limitations of physical sensors is critical.

- **Realism vs. Speed:** Balancing the level of realism in simulations with the computational efficiency required for testing large-scale systems.

- **Validation:** Ensuring that simulation results align with real-world performance to instill confidence in the technology.

6. Future Trends and Innovations:

The future of robot simulation and testing involves advancements in machine learning for generating more realistic environments and AI-driven testing frameworks for assessing robot behavior in complex, dynamic scenarios.

In conclusion, robot simulation and testing are indispensable stages in the development and validation of autonomous machines. As technology continues to evolve, these virtual proving grounds will play an increasingly crucial role in ensuring the safety, reliability, and efficacy of robotic systems across various industries and applications.

E. Automation Implementation in Business Processes (BPA): Streamlining Operations for Efficiency and Agility

Automation implementation in business processes, often referred to as Business Process Automation (BPA), is a strategic approach that leverages technology to optimize and streamline workflows. BPA aims to reduce manual intervention, enhance accuracy, and boost operational efficiency across various sectors and industries. In this in-depth exploration, we will delve into the intricacies of automation implementation in business processes, examining its methodologies, applications, challenges, and the transformative impact it has on organizations striving for greater agility and competitiveness.

1. Significance of Automation Implementation in Business Processes:

- **Efficiency:** BPA eliminates repetitive, time-consuming tasks, allowing employees to focus on more valuable, strategic activities.

- **Accuracy:** Automation reduces human errors, enhancing the quality and consistency of work.

- **Scalability:** Automated processes can handle increased workloads without a proportional increase in labor.

- **Compliance:** BPA ensures adherence to regulatory requirements by enforcing standardized processes and audit trails.

2. Key Components of Automation Implementation in BPA:

- **Process Mapping:** Identifying, analyzing, and documenting current business processes to identify automation opportunities.

- **Automation Tools:** Utilizing software and technologies like workflow automation platforms, robotic process automation (RPA) bots, and AI-driven solutions.

- **Integration:** Seamlessly integrating automation solutions with existing IT infrastructure and applications.

- **Monitoring and Analytics:** Implementing tools to monitor process performance and gather data for continuous improvement.

3. Methodologies and Approaches in Automation Implementation:

- **Robotic Process Automation (RPA):** Deploying software bots to mimic human actions and automate rule-based, repetitive tasks.

- **Artificial Intelligence (AI):** Using machine learning and natural language processing to automate tasks that require cognitive abilities, such as data analysis and decision-making.

- **Workflow Automation:** Designing and automating entire end-to-end business processes, often involving multiple steps and participants.

4. Applications of Automation Implementation in BPA:

- **Finance and Accounting:** Automating invoice processing, expense management, and financial reporting.

- **Human Resources:** Streamlining recruitment, onboarding, and payroll processes.

- **Customer Service:** Automating customer inquiries, support tickets, and chatbots for enhanced user experiences.

- **Supply Chain:** Optimizing order processing, inventory management, and demand forecasting.

- **Healthcare:** Automating patient records management, appointment scheduling, and billing.

5. Challenges in Automation Implementation in BPA:

- **Resistance to Change:** Employees may resist automation due to fears of job displacement or unfamiliarity with new tools.

- **Complexity:** Integrating automation into complex, legacy systems and processes can be challenging.

- **Data Security:** Ensuring that sensitive data remains secure and compliant with privacy regulations.

- **ROI Measurement:** Determining the return on investment (ROI) of automation initiatives can be complex and time-consuming.

6. Future Trends and Innovations:

The future of automation implementation in BPA involves the increasing use of AI-driven decision-making, low-code/no-code development platforms for rapid automation, and greater emphasis on intelligent automation that combines RPA, AI, and machine learning.

In conclusion, automation implementation in business processes is a strategic imperative for organizations seeking to enhance efficiency, reduce costs, and stay competitive in a rapidly evolving business landscape. As technology continues to advance, BPA will play an increasingly pivotal role in reshaping the way organizations operate and deliver value to customers.

CHAPTER 8

Robot Collaboration, Swarm Robotics, and Home Automation: The Synergy of Intelligent Agents in Everyday Life

Robot collaboration, swarm robotics, and home automation represent a convergence of technologies that bring intelligence and automation into our daily lives. These paradigms harness the power of multiple intelligent agents, whether physical robots or interconnected devices, to work together efficiently and seamlessly. In this introductory exploration, we embark on a journey into the world of robot collaboration, swarm robotics, and home automation, uncovering their significance, applications, and the transformative potential they hold for revolutionizing how we live, work, and interact with technology in our homes and beyond.

A. Multi-Robot Systems and Cooperative Behavior: The Symphony of Coordination

Multi-robot systems and cooperative behavior represent a transformative approach to robotics, where multiple intelligent agents collaborate seamlessly to achieve complex tasks. This paradigm shift from solitary robots to coordinated teams has far-reaching implications across various domains, from manufacturing and search and rescue to environmental monitoring and space exploration. In this in-depth exploration, we delve into the intricacies of multi-robot systems and cooperative behavior, examining their methodologies, applications, challenges, and the transformative impact they have on reshaping the capabilities and efficiency of robotic systems.

1. Significance of Multi-Robot Systems and Cooperative Behavior:

- **Enhanced Efficiency:** Coordinated robots can accomplish tasks more efficiently, rapidly, and with greater precision than individual robots.

- **Robustness:** Redundancy in multi-robot systems allows them to adapt and continue functioning even if some robots fail or are removed from the team.

- **Scalability:** Additional robots can be added to the team to handle larger workloads or more extensive environments.

- **Versatility:** Cooperative robots can tackle a wide range of tasks, from collaborative manipulation to exploring hazardous or unknown environments.

2. Key Components of Multi-Robot Systems:

- **Communication Infrastructure:** Robots exchange information through wireless networks, enabling coordination and collaboration.

- **Control Algorithms:** Algorithms govern the behavior and decision-making of individual robots and the team as a whole.

- **Sensing and Perception:** Sensors, such as cameras, LiDAR, and proximity sensors, provide data on the robot's environment and its teammates.

- **Task Allocation:** Mechanisms for distributing tasks among robots, ensuring efficient resource utilization.

3. Methodologies and Approaches in Multi-Robot Systems:

- **Swarm Robotics:** Inspired by natural swarms, this approach emphasizes simplicity and local interactions among robots, leading to emergent, collective behaviors.

- **Centralized vs. Decentralized Control:** Multi-robot systems can be centrally controlled, with one entity making decisions for the entire team, or decentralized, where each robot contributes to decision-making.

- **Cooperative Game Theory:** Mathematical models and strategies from game theory help optimize cooperation and resource sharing among robots.

4. Applications of Multi-Robot Systems and Cooperative Behavior:

- **Manufacturing:** In manufacturing plants, teams of robots collaborate on tasks like assembly, welding, and quality control.

- **Search and Rescue:** In disaster scenarios, robot teams can search for survivors, deliver supplies, and map hazardous environments.

- **Agriculture:** Cooperative robots are used in precision agriculture for tasks like planting, harvesting, and pest control.

- **Space Exploration:** Robotic teams explore planets and moons, where communication delays make real-time human control impossible.

- **Environmental Monitoring:** Robot teams monitor and collect data from remote or hazardous environments, such as forests, oceans, and nuclear facilities.

5. Challenges in Multi-Robot Systems and Cooperative Behavior:

- **Communication:** Ensuring reliable communication among robots, especially in dynamic and noisy environments.

- **Coordination:** Robots must coordinate their actions

effectively, avoiding collisions and conflicts while achieving the task's objectives.

- **Scalability:** As the number of robots increases, managing their interactions and preventing congestion becomes increasingly complex.

- **Resource Allocation:** Determining how to allocate tasks and resources optimally among robots in a cooperative manner.

6. Future Trends and Innovations:

The future of multi-robot systems involves integrating advanced AI and machine learning techniques to enable robots to learn from each other, adapt to dynamic environments, and exhibit greater autonomy in decision-making.

In conclusion, multi-robot systems and cooperative behavior are reshaping the landscape of robotics, enabling teams of robots to collaborate effectively in diverse applications. As these technologies continue to advance, we can expect to see even more sophisticated and versatile robotic teams that play a pivotal role in addressing complex challenges across industries and domains.

B. Swarm Intelligence in Robotics: The Collective Wisdom of Machines

Swarm intelligence in robotics is a fascinating paradigm

inspired by the behavior of social insects and other organisms that exhibit collective intelligence. It involves coordinating large groups of relatively simple robots or agents to solve complex tasks. This approach has garnered significant attention due to its potential applications in various domains, from search and rescue missions to environmental monitoring and autonomous transportation. In this in-depth exploration, we delve into the intricacies of swarm intelligence in robotics, examining its methodologies, applications, challenges, and the transformative impact it has on creating adaptive and efficient robotic systems.

1. Significance of Swarm Intelligence in Robotics:

- **Robustness:** Swarm robots can continue functioning even if some members fail, making them resilient in challenging environments.

- **Scalability:** As the number of robots increases, swarm systems can adapt to handle larger tasks or environments without a proportional increase in complexity.

- **Flexibility:** Swarm robots are versatile and can be deployed for a wide range of tasks, including exploration, surveillance, and monitoring.

- **Distributed Sensing:** Swarm robots can collectively gather and process data from multiple sensors, enabling comprehensive environmental awareness.

2. Key Concepts in Swarm Intelligence:

- **Emergence:** Complex behaviors and patterns emerge from simple interactions among individual agents. No central control is required.

- **Local Interaction:** Swarm robots make decisions based on information from nearby robots, rather than relying on global knowledge.

- **Self-Organization:** The system as a whole self-organizes to achieve goals, even when individual agents follow simple rules.

- **Stigmergy:** Communication occurs indirectly through the environment, where robots leave traces (pheromones) that influence the behavior of others.

3. Methodologies and Approaches in Swarm Intelligence:

- **Particle Swarm Optimization (PSO):** Inspired by the flocking behavior of birds, PSO is an optimization technique where a swarm of particles iteratively adjusts their positions to find optimal solutions.

- **Ant Colony Optimization (ACO):** Modeled after the foraging behavior of ants, ACO algorithms find optimal paths or solutions by simulating the pheromone-based communication among ants.

- **Swarm Robotics Algorithms:** These algorithms govern the behavior and interactions of individual robots in a swarm, allowing them to collectively achieve objectives, such as exploration or coverage.

4. Applications of Swarm Intelligence in Robotics:

- **Search and Rescue:** Swarms of robots can search for survivors in disaster-stricken areas, improving the chances of finding individuals in need of assistance.

- **Environmental Monitoring:** Swarm robots can be deployed to collect data on pollution levels, wildlife tracking, and ecosystem analysis.

- **Agriculture:** Swarm robots can work collaboratively to perform tasks like crop monitoring, planting, and harvesting.

- **Transportation:** Autonomous vehicles can use swarm intelligence to coordinate traffic and optimize routes, reducing congestion and energy consumption.

5. Challenges in Swarm Intelligence in Robotics:

- **Scalability:** Ensuring that swarm algorithms remain efficient and effective as the number of robots in the swarm increases.

- **Robustness:** Managing the robustness of swarm systems in unpredictable and dynamic environments.

- **Communication:** Designing efficient communication mechanisms for swarm robots, particularly in scenarios with limited bandwidth or interference.

- **Resource Management:** Allocating tasks and resources optimally within the swarm to achieve mission objectives.

6. Future Trends and Innovations:

The future of swarm intelligence in robotics involves the integration of advanced machine learning techniques, allowing swarm robots to learn from their experiences and adapt to new challenges autonomously.

In conclusion, swarm intelligence in robotics offers a powerful approach to addressing complex problems by harnessing the collective wisdom of multiple robots. As technology continues to advance, we can expect to see increasingly sophisticated swarm systems that have a profound impact on various industries and applications, from disaster response to environmental conservation.

C. Distributed Robotics: Unleashing the Power of Cooperative Autonomy

Distributed robotics represents a transformative paradigm in which a group of autonomous robotic agents collaborates to achieve common goals. Unlike traditional robotics, where a single, centralized robot performs tasks, distributed robotics harnesses the power of collective intelligence and teamwork. This approach has the potential to revolutionize various fields, from environmental monitoring and exploration to healthcare and disaster response. In this in-depth exploration, we delve into the intricacies of distributed robotics, examining its methodologies, applications, challenges, and the transformative impact it has on creating adaptable, scalable, and versatile robotic systems.

1. Significance of Distributed Robotics:

- **Scalability:** Distributed systems can easily scale up by adding more agents to tackle larger tasks or cover larger areas.

- **Robustness:** Redundancy in the form of multiple robots ensures that the system can continue to function even if some agents fail.

- **Flexibility:** Distributed robots are versatile and can adapt to a wide range of tasks and environments.

- **Resource Efficiency:** Cooperative agents can optimize resource usage and share the computational burden, making them more efficient.

2. Key Concepts in Distributed Robotics:

- **Decentralization:** Distributed robots operate without central control, making decisions autonomously based on local information and interactions.

- **Communication:** Agents communicate with each other to share information, coordinate actions, and achieve common goals.

- **Emergent Behavior:** Complex, collective behaviors emerge from the interactions and decisions of individual agents, often without centralized planning.

- **Collaborative Autonomy:** Robots cooperate to perform tasks that are challenging or impossible for a single robot to accomplish.

3. Methodologies and Approaches in Distributed Robotics:

- **Swarm Robotics:** Inspired by the behavior of social insects, swarm robotics emphasizes simple rules and local interactions among robots to achieve complex tasks.

- **Multi-Agent Systems:** Involves designing intelligent agents that can reason, plan, and make decisions independently while collaborating with others.

- **Decentralized Control:** Each robot has its own control and decision-making capabilities, and coordination emerges from local interactions.

- **Communication Protocols:** Develops efficient

communication protocols for robots to exchange information and coordinate their actions.

4. Applications of Distributed Robotics:

- **Environmental Monitoring:** Distributed robot teams can monitor wildlife, track weather patterns, and collect data on environmental changes.

- **Search and Rescue:** In disaster scenarios, multiple robots can work together to search for survivors and navigate hazardous environments.

- **Agriculture:** Distributed robots can perform tasks like planting, harvesting, and pest control, improving crop yield and resource efficiency.

- **Space Exploration:** In planetary exploration, distributed robots can work together to explore diverse terrains and conduct scientific experiments.

5. Challenges in Distributed Robotics:

- **Communication Reliability:** Ensuring robust communication among agents in dynamic and noisy environments.

- **Coordination:** Coordinating the actions of multiple agents to achieve common objectives efficiently.

- **Scalability:** Managing the complexity and scalability of distributed systems as the number of agents increases.

- **Resource Allocation:** Allocating tasks and resources optimally within the distributed system to achieve mission objectives.

6. Future Trends and Innovations:

The future of distributed robotics involves advancements in machine learning, AI, and communication technologies, enabling robots to adapt, learn from experience, and autonomously coordinate in complex and dynamic environments.

In conclusion, distributed robotics represents a transformative approach to robotics, harnessing the power of collective autonomy and collaboration among multiple agents. As this field continues to advance, we can anticipate increasingly sophisticated and versatile robotic systems that play pivotal roles in addressing complex challenges across a wide range of industries and applications.

D. Home Automation Systems: Transforming Houses into Smart Homes

Home automation systems have redefined the way we interact with our living spaces, making our homes more convenient, energy-efficient, and secure. These systems leverage cutting-edge technologies to connect and automate various devices and services, allowing homeowners to control and monitor their environments with ease. In this in-depth exploration, we delve into the intricacies of home automation systems, examining their methodologies, applications, challenges, and the transformative impact they have on modern living.

1. Significance of Home Automation Systems:

- **Convenience:** Automation allows homeowners to control lighting, heating, cooling, security, and entertainment systems with the touch of a button or voice command.

- **Energy Efficiency:** Smart thermostats, lighting, and appliances help reduce energy consumption, leading to cost savings and environmental benefits.

- **Security:** Surveillance cameras, smart locks, and motion detectors enhance home security by providing real-time monitoring and alerts.

- **Accessibility:** Automation systems can be customized to accommodate individuals with disabilities, making homes more inclusive.

2. Key Components of Home Automation Systems:

- **Sensors:** Devices like motion sensors, temperature sensors, and light sensors collect data from the environment.

- **Actuators:** These components, such as motors or switches, carry out actions based on commands from the automation system.

- **Control Hub:** The central unit that connects and manages all devices, often controlled through a smartphone app or voice assistant.

- **Communication Protocols:** Wireless technologies like Wi-Fi, Zigbee, Z-Wave, or Bluetooth enable devices to communicate with each other and the central hub.

3. Methodologies and Approaches in Home Automation Systems:

- **Rule-Based Automation:** Users set specific rules or conditions for devices to follow, such as turning on lights when motion is detected.

- **Machine Learning and AI:** Systems can learn user preferences and adapt automation based on historical data and patterns.

- **Voice Control:** Integration with voice assistants like Amazon Alexa or Google Assistant allows for intuitive control.

4. Applications of Home Automation Systems:

- **Lighting Control:** Smart lighting systems enable users to adjust brightness, color, and schedule lighting based on their preferences.

- **Climate Control:** Smart thermostats optimize heating and cooling, saving energy without sacrificing comfort.

- **Security and Surveillance:** Automation systems provide real-time alerts and remote monitoring, enhancing home security.

- **Entertainment:** Users can automate audio and video systems, creating immersive home theater experiences.

- **Appliance Control:** Smart appliances can be remotely controlled, making chores more efficient and flexible.

5. Challenges in Home Automation Systems:

- **Interoperability:** Ensuring that devices from different manufacturers can communicate and work together seamlessly.

- **Security Concerns:** Protecting home automation systems from cyberattacks and unauthorized access.

- **Complexity:** Managing and configuring a growing number of devices and automation rules can be overwhelming for users.

- **Privacy:** The collection of data by automation systems raises privacy concerns, necessitating robust data protection measures.

6. Future Trends and Innovations:

The future of home automation systems will likely involve increased integration with AI and machine learning, leading to more context-aware and anticipatory automation. Additionally, edge computing will enhance device responsiveness and reduce dependence on cloud services.

In conclusion, home automation systems have reshaped our homes, offering unparalleled convenience, energy efficiency, and security. As technology continues to evolve, these systems will become even more sophisticated, enhancing our daily lives and making our homes smarter and more adaptable to our needs.

E. IoT Automation and Robotics: A Synergistic Leap into the Future

IoT (Internet of Things) automation and robotics represent a powerful convergence of technologies that promises to revolutionize industries and daily life. This amalgamation involves the integration of IoT devices, sensors, and robotics to create intelligent, autonomous systems capable of diverse tasks. From manufacturing and healthcare to smart homes and agriculture, IoT automation and robotics are driving innovation and efficiency. In this in-depth exploration, we delve into the intricacies of this transformative field, examining its methodologies, applications, challenges, and the profound impact it has on reshaping industries and the way we interact with technology.

1. Significance of IoT Automation and Robotics:

- **Efficiency:** IoT automation enhances operational efficiency by enabling real-time data analysis and informed decision-making.

- **Productivity:** Automation and robotics can perform repetitive tasks tirelessly, leading to increased productivity and cost savings.

- **Safety:** In hazardous environments, robots can perform tasks, minimizing human exposure to risks.

- **Quality Control:** IoT sensors and robotics together ensure consistent and high-quality production processes.

2. Key Components of IoT Automation and Robotics:

- **IoT Devices:** Sensors, actuators, and connected devices collect data and transmit it over the internet.

- **Robotic Systems:** Robots, drones, and autonomous vehicles execute tasks based on data and commands received.

- **Communication Networks:** High-speed, reliable networks

enable seamless data exchange between devices and central systems.

- **Cloud Computing:** IoT data is often processed, analyzed, and stored in the cloud, providing scalability and accessibility.

3. Methodologies and Approaches in IoT Automation and Robotics:

- **Data Analytics:** Advanced analytics and machine learning algorithms process IoT data to derive insights and make autonomous decisions.

- **Edge Computing:** Local processing at the device level enhances real-time response and reduces latency in critical applications.

- **Human-Robot Collaboration:** Robots and humans work together safely in a variety of settings, from manufacturing to healthcare.

4. Applications of IoT Automation and Robotics:

- **Manufacturing:** IoT automation and robotics optimize production lines, quality control, and predictive maintenance.

- **Healthcare:** Robotics assist in surgeries, IoT devices monitor patients' vital signs, and automated drug dispensing improves efficiency.

- **Agriculture:** IoT sensors gather data on soil conditions, crop health, and weather, while robots perform tasks like harvesting and weeding.

- **Logistics and Transportation:** Autonomous vehicles and drones streamline delivery and transport operations.

- **Smart Homes:** IoT automation manages lighting, heating, security, and appliances for enhanced convenience and energy efficiency.

5. Challenges in IoT Automation and Robotics:

- **Security:** Protecting IoT systems from cyberattacks, data breaches, and unauthorized access is paramount.

- **Interoperability:** Ensuring that IoT devices from various manufacturers can communicate and function together seamlessly.

- **Data Privacy:** Managing and securing sensitive data collected by IoT devices and robots.

- **Regulatory Compliance:** Adhering to evolving regulations and standards in different industries.

6. Future Trends and Innovations:

The future of IoT automation and robotics lies in greater autonomy and adaptability, with robots learning from data and improving their performance over time. Edge AI and 5G networks will facilitate faster, more responsive IoT systems.

In conclusion, IoT automation and robotics are poised to transform industries and our daily lives. As technology continues to advance, the synergy between IoT devices and robotics will enable smarter, more efficient, and safer automation systems, paving the way for a future characterized by unprecedented connectivity and automation.

CHAPTER 9

Robot Ethics, Safety, and Regulations in Automation: Navigating the Moral and Regulatory Landscape

Robot ethics, safety, and regulations are central to the responsible development and deployment of automation technologies. As robots and automated systems become increasingly integrated into our lives, it is essential to consider the ethical implications, ensure safety, and establish regulatory frameworks that govern their use. In this introductory exploration, we embark on a journey into the realm of robot ethics, safety, and regulations, shedding light on the complex interplay of moral considerations, safety measures, and legal constraints that shape the future of automation and robotics.

A. Robot Safety Considerations: Ensuring the Well-being of Humans and Machines

Robot safety considerations are paramount in the design, deployment, and operation of robotic systems. As robots become more prevalent in various industries and settings, ensuring the safety of both humans and machines is essential. This comprehensive exploration delves into the intricacies of robot safety, examining its significance, principles, methods, and the evolving landscape of safety standards and practices.

1. Significance of Robot Safety:

- **Human Well-being:** Robot safety is crucial to protect human operators, bystanders, and anyone who interacts with robotic systems.

- **Cost Savings:** Ensuring safety reduces the risk of accidents, damage to equipment, and costly legal liabilities.

- **Productivity:** Safe robots can operate in close proximity to humans, improving productivity and collaboration in various industries.

- **Public Trust:** Safe robotic technologies foster public trust and acceptance, enabling their broader adoption.

2. Key Principles of Robot Safety:

- **Risk Assessment:** Identifying potential hazards and assessing the associated risks is the foundation of robot safety.

- **Hazard Mitigation:** Implementing engineering controls, such as safety barriers or emergency stop mechanisms, to mitigate identified hazards.

- **Safety Protocols:** Establishing clear safety protocols and procedures for human-robot interactions, including training and safe work practices.

- **Fail-Safe Mechanisms:** Incorporating fail-safe mechanisms that enable robots to respond to faults or emergencies by halting or entering a safe state.

3. Methods and Approaches in Robot Safety:

- **Risk Assessment Tools:** Software tools and methodologies help evaluate the safety of robotic systems by identifying potential risks.

- **Collaborative Robotics:** Collaborative robots (cobots) are designed to work safely alongside humans, often with features like force-limiting technology and vision systems.

- **Sensors and Perception:** Robots equipped with sensors, such as cameras and LiDAR, can detect and respond to changes in their environment, avoiding collisions.

- **Machine Learning:** AI and machine learning algorithms can enhance safety by enabling robots to learn and adapt to their surroundings and avoid risky behaviors.

4. Evolving Safety Standards and Regulations:

- **ISO Standards:** The International Organization for Standardization (ISO) has developed a series of standards (e.g., ISO 10218 and ISO 13482) that provide guidelines for robot safety.

- **ANSI/RIA Standards:** The American National Standards Institute (ANSI) and the Robotic Industries Association (RIA) in the United States contribute to developing safety standards for robotics.

- **National and Regional Regulations:** Various countries and regions have specific regulations and guidelines governing robot safety, which may vary in scope and stringency.

5. Challenges in Robot Safety:

- **Human-Robot Interaction:** Ensuring safety in scenarios where humans and robots work closely together can be challenging due to unpredictable human behavior.

- **Complexity:** As robots become more autonomous and capable, ensuring their safety becomes increasingly complex.

- **Updating Standards:** The rapid pace of technological advancement necessitates the continuous updating of safety standards and regulations.

- **Ethical Considerations:** Safety extends to ethical concerns, such as the potential for biased AI algorithms or the ethical use of autonomous systems in decision-making.

6. Future Trends and Innovations:

The future of robot safety involves the development of adaptive safety systems that can dynamically respond to changing conditions and the integration of safety considerations into AI and machine learning algorithms.

In conclusion, robot safety considerations are foundational to the responsible adoption of robotic technologies across industries. As robotics continues to evolve, so too will the methods, standards, and practices that ensure the safety of robots, humans, and society at large.

B. Ethical Issues in Robotics and Automation: Navigating the Moral Landscape of AI and Machines

Ethical issues in robotics and automation have become increasingly prominent as these technologies play a more significant role in our lives. As robots and automated systems become more sophisticated and autonomous, they raise complex moral questions that touch on human values, societal norms, and the ethical implications of creating intelligent machines. In this in-depth exploration, we delve into the multifaceted ethical challenges posed by robotics and automation, examining their significance, key concerns, and potential solutions.

1. Significance of Ethical Issues in Robotics and Automation:

- **Human Impact:** Robotics and automation directly impact human lives, well-being, and livelihoods, making ethical considerations paramount.

- **Autonomy:** As robots and AI systems become more autonomous, they raise questions about accountability, decision-making, and the potential for unintended consequences.

- **Societal Values:** Ethical issues in this field touch on fundamental societal values, including privacy, fairness, safety, and transparency.

- **Public Trust:** Addressing ethical concerns fosters public trust in these technologies, which is essential for their widespread adoption and acceptance.

2. Key Ethical Concerns in Robotics and Automation:

- **Job Displacement:** Automation can lead to job loss and economic disruption, raising questions about the societal responsibility to address these impacts.

- **Bias and Discrimination:** AI algorithms can perpetuate bias and discrimination when making decisions related to hiring, lending, criminal justice, and more.

- **Privacy:** The collection and use of data by robots and automated systems raise concerns about personal privacy and data security.

- **Autonomous Decision-Making:** As robots make decisions independently, there is a need to ensure that their choices align with human values and ethical principles.

- **Lethal Autonomous Weapons:** The development of autonomous military robots raises ethical concerns about the use of lethal force without human intervention.

3. Ethical Frameworks and Approaches:

- **Utilitarianism:** Focuses on the greatest good for the greatest number and seeks to maximize overall happiness and well-being.

- **Deontology:** Emphasizes adherence to moral principles, irrespective of the consequences, and often invokes concepts like rights, duties, and categorical imperatives.

- **Virtue Ethics:** Considers the moral character of individuals and aims to cultivate virtues such as honesty, compassion, and integrity.

- **Ethics of Care:** Prioritizes relationships and empathy, emphasizing the importance of caring for others and addressing their needs.

- **Transparency and Accountability:** Ethical practices include ensuring transparency in algorithms, decision-making processes, and holding creators accountable for the behavior of their creations.

4. Ethical Decision-Making in Robotics and AI:

- **Design Ethics:** Ethical considerations should be integrated into the design and development phase of robotics and AI systems.

- **Roboethics:** The emerging field of roboethics examines the ethical implications of robotics and AI technologies and aims to guide their responsible development and use.

- **Ethical Audits:** Regular ethical audits and assessments can help identify and rectify potential ethical issues in existing systems.

- **Public Engagement:** Including a diverse range of stakeholders in ethical discussions and decision-making processes ensures a more comprehensive perspective.

5. Challenges in Addressing Ethical Issues:

- **Complexity:** Ethical considerations in robotics and AI are often multifaceted, requiring careful analysis and nuanced solutions.

- **Rapid Advancement:** The pace of technological advancement can outstrip the development of ethical frameworks and regulations.

- **Global Consensus:** Achieving consensus on ethical standards and regulations across different regions and cultures is challenging.

- **Unintended Consequences:** Ethical decisions made during design may not anticipate all possible consequences or scenarios.

6. Future Trends and Innovations:

The future of ethical considerations in robotics and automation will likely involve the development of AI systems that are more transparent, explainable, and capable of aligning their behavior with human values.

In conclusion, addressing ethical issues in robotics and automation is essential for the responsible and sustainable development of these technologies. As they continue to evolve and permeate various aspects of society, the ethical dimension will remain central to shaping their impact on individuals, communities, and the world at large.

C. Regulations and Standards for Robotics: Ensuring Safety, Quality, and Compliance

Regulations and standards play a pivotal role in shaping the development, deployment, and safe operation of robotic systems. They provide essential guidelines and requirements that promote uniformity, safety, and interoperability in the field of robotics. This in-depth exploration delves into the world of regulations and standards for robotics, examining their significance, key players, primary objectives, and their impact on the industry and society.

1. Significance of Regulations and Standards:

- **Safety Assurance:** Regulations and standards establish safety measures to protect humans and ensure the safe operation of robots.

- **Quality Assurance:** Standards promote quality control and consistency in the design and manufacturing of robotic systems.

- **Interoperability:** Standards enable different robotic systems to work together seamlessly, facilitating compatibility and integration.

- **Global Trade:** Common international standards facilitate the global trade of robotic products and services.

2. Key Players in Establishing Standards:

- **ISO (International Organization for Standardization):** ISO is a globally recognized body that develops international standards for various industries, including robotics. ISO standards are widely adopted and respected worldwide.

- **ANSI (American National Standards Institute):** ANSI oversees the development of standards in the United States, including those related to robotics. ANSI standards often align with ISO standards.

- **IEC (International Electrotechnical Commission):** IEC focuses on the standardization of electrical and electronic technologies, including those used in robotics.

- **IEEE (Institute of Electrical and Electronics Engineers):** IEEE develops technical standards for various fields, including robotics and automation.

3. Primary Objectives of Regulations and Standards:

- **Safety:** Ensuring that robotic systems do not pose undue risks to humans, property, or the environment.

- **Quality Control:** Defining requirements for the design, manufacturing, and performance of robotic systems to guarantee their quality and reliability.

- **Interoperability:** Establishing common communication protocols and interfaces to enable different robotic systems to work together effectively.

- **Ethics and Governance:** Addressing ethical considerations, such as the use of autonomous robots in decision-making, and providing guidance on responsible robotic development and deployment.

4. Key Robotics Standards and Regulations:

- **ISO 10218:** Provides safety requirements for industrial robots, including guidelines for the design and integration of safety features.

- **ISO 13482:** Focuses on the safety of personal care robots, setting safety requirements for robots intended to assist humans.

- **ISO 18646:** Defines the terms and concepts related to service robots, promoting consistency in terminology.

- **ANSI/RIA R15.06:** Establishes safety standards for industrial robots in the United States, often aligning with ISO standards.

- **IEC 61508:** Provides a framework for the functional safety of electrical, electronic, and programmable electronic systems, which includes many robotic systems.

5. Challenges in Robotics Regulations and Standards:

- **Rapid Technological Advancements:** Keeping standards up-to-date with the rapid evolution of robotics technology is challenging.

- **Global Consensus:** Achieving consensus on standards across diverse regions and industries can be complex.

- **Ethical and Social Considerations:** Standards may need to address ethical and societal concerns, such as the use of robots in healthcare and decision-making.

- **Interdisciplinary Nature:** Robotics is an interdisciplinary field, and standards often need to integrate aspects of mechanics, electronics, and software.

6. Future Trends and Innovations:

The future of robotics regulations and standards will likely involve increased emphasis on safety in collaborative robotics (cobots), ethical considerations in autonomous systems, and the development of international standards to facilitate global robotic deployment.

In conclusion, regulations and standards are fundamental pillars that ensure the responsible and safe development of robotics. They not only provide guidance for manufacturers and operators but also contribute to the trust and acceptance of robotic systems in society. As robotics technology continues to advance, the role of regulations and standards will remain crucial in shaping the industry's future.

D. Liability and Responsibility in Robotics and Automation: Navigating the Complex Legal Landscape

The rise of robotics and automation technologies has ushered in a new era of innovation and efficiency across industries. However, this transformation also brings complex legal and ethical questions about liability and responsibility. As machines become more autonomous and capable of making decisions, it becomes essential to establish frameworks that determine who is liable for actions or harm caused by robotic systems. This comprehensive exploration delves into the intricate realm of liability and responsibility in robotics and automation, examining its significance, key challenges, legal considerations, and potential solutions.

1. Significance of Liability and Responsibility:

- **Accountability:** Determining liability and responsibility is crucial to hold parties accountable for actions or decisions made by autonomous robots.

- **Risk Mitigation:** Assigning liability helps incentivize manufacturers and operators to prioritize safety in robotic design and deployment.

- **Legal Clarity:** Establishing legal frameworks for liability clarifies expectations and potential consequences in case of accidents or harm.

- **Ethical Considerations:** Responsibility in robotics extends to ethical considerations, such as ensuring the ethical use of AI and autonomous systems.

2. Key Challenges in Determining Liability:

- **Autonomy Levels:** Robots and automated systems vary in their levels of autonomy, making it challenging to define responsibility when a machine operates independently.

- **Human-Machine Interaction:** Determining liability becomes complex when humans interact with robots, particularly in collaborative settings.

- **Causality and Attribution:** Identifying the direct cause of an incident involving autonomous systems can be challenging, affecting liability determination.

- **Insurance and Coverage:** The insurance industry is still adapting to the unique risks posed by robotics and automation, impacting liability considerations.

3. Legal Considerations and Approaches:

- **Product Liability:** Traditional product liability laws may be applied to hold manufacturers responsible for defects or malfunctions in robots.

- **Strict Liability:** Some jurisdictions impose strict liability on manufacturers, making them liable for any harm caused by their products, regardless of fault.

- **Contractual Agreements:** Parties involved in robotic deployment may establish contractual agreements that allocate responsibility.

- **Regulatory Frameworks:** Governments and regulatory bodies may introduce specific regulations and liability frameworks for robotics and automation.

4. International and Regional Perspectives:

- **EU Robot Law Proposal:** The European Union has considered the introduction of robot-specific regulations, including rules on liability for harm caused by robots.

- **Uniform Commercial Code (UCC) in the U.S.:** The UCC includes provisions on product liability, which can apply to robots and automation systems.

- **International Treaty:** Some experts advocate for the creation of an international treaty that establishes liability rules for autonomous systems.

5. Future Trends and Innovations:

The future of liability and responsibility in robotics and automation will likely involve more sophisticated legal frameworks that adapt to evolving technology. These frameworks may include AI's ability to explain its decisions, ethical considerations, and the development of autonomous legal entities.

In conclusion, liability and responsibility in robotics and automation are critical aspects that require careful consideration as technology continues to advance. Striking a balance between promoting innovation and ensuring accountability is essential to harness the benefits of automation while mitigating risks and protecting individuals and society as a whole.

CHAPTER 10

Robotics and Automation Applications: Transforming Industries and Daily Life

The realm of robotics and automation applications is characterized by its profound impact on a wide spectrum of industries and everyday experiences. From revolutionizing manufacturing processes to enhancing healthcare, agriculture, and our homes, these technologies have become ubiquitous, driving efficiency, safety, and innovation. In this introductory exploration, we embark on a journey into the vast landscape of robotics and automation applications, shedding light on their significance, diverse use cases, and the transformative influence they exert across multiple sectors.

A. Industrial Robotics and Automation: Revolutionizing Manufacturing and Beyond

Industrial robotics and automation represent a pivotal technological leap that has reshaped manufacturing and various industries, driving efficiency, precision, and competitiveness. These systems, comprised of robotic arms, sensors, and intelligent control, have become integral to modern production processes. In this in-depth exploration, we delve into the world of industrial robotics and automation, examining their significance, key components, applications, challenges, and the profound impact they have on manufacturing and beyond.

1. Significance of Industrial Robotics and Automation:

- **Enhanced Efficiency:** Automation streamlines production, reduces cycle times, and enhances productivity by replacing repetitive manual tasks with robotic precision.

- **Quality Assurance:** Robots ensure consistency and precision in manufacturing, leading to higher-quality products with fewer defects.

- **Cost Reduction:** Automation reduces labor costs, minimizes material wastage, and optimizes resource utilization, contributing to cost savings.

- **Safety Improvement:** Dangerous and strenuous tasks can be automated, improving workplace safety and reducing the risk of injuries.

2. Key Components of Industrial Robotics and Automation:

- **Robotic Arms:** These mechanical arms, equipped with various end-effectors, perform tasks such as welding, assembly, and material handling.

- **Sensors:** Proximity sensors, vision systems, and force sensors provide robots with real-time feedback and environmental awareness.

- **Control Systems:** Advanced control algorithms and software orchestrate the actions of robots, ensuring precision and adaptability.

- **Programming Interfaces:** Intuitive programming interfaces enable human operators to teach and configure robots for specific tasks.

3. Applications of Industrial Robotics and Automation:

- **Manufacturing:** Robots are widely used in automotive, electronics, aerospace, and consumer goods industries for assembly, welding, painting, and more.

- **Material Handling:** Automation systems transport and manage materials within factories, distribution centers, and warehouses.

- **Quality Control:** Vision systems and sensors inspect products for defects, ensuring adherence to quality standards.

- **Packaging and Palletizing:** Robots are employed in packaging operations, from sorting and filling to sealing and palletizing.

- **Heavy Industry:** In industries such as mining and construction, robots handle hazardous or heavy-duty tasks, improving safety and efficiency.

4. Challenges in Industrial Robotics and Automation:

- **Integration Complexity:** Integrating automation into existing production lines can be complex and require specialized expertise.

- **Costs:** Initial setup costs and ongoing maintenance expenses can be significant, especially for small and medium-sized enterprises.

- **Adaptability:** Robots may struggle with tasks that require fine dexterity or adaptability in unstructured environments.

- **Human Collaboration:** Ensuring safe human-robot collaboration in shared workspaces requires advanced safety measures.

5. Future Trends and Innovations:

The future of industrial robotics and automation involves greater flexibility, adaptability, and collaboration with humans. Advances in AI, machine learning, and cloud computing are expected to enhance robotic capabilities, enabling them to handle more complex tasks and adapt to changing production demands.

In conclusion, industrial robotics and automation have not only revolutionized manufacturing but also laid the foundation for increased productivity and competitiveness across industries. As technology continues to advance, these systems will play an even more significant role in shaping the future of production and expanding their applications to new domains.

B. Healthcare Robotics: Revolutionizing Patient Care and Medical Practices

Healthcare robotics is a burgeoning field that has the potential to transform the medical landscape, offering innovative solutions to enhance patient care, improve clinical workflows, and address healthcare challenges. With the integration of robotics and automation, healthcare professionals can perform tasks more efficiently, accurately, and safely. In this comprehensive exploration, we delve into the world of healthcare robotics, examining its significance, key applications, benefits, challenges, and the promising future it holds for the healthcare industry.

1. Significance of Healthcare Robotics:

- **Precision and Accuracy:** Robots can perform precise surgical procedures and deliver medications with exceptional accuracy, minimizing human errors.

- **Enhanced Efficiency:** Automation in healthcare streamlines administrative tasks, reducing paperwork, and allowing medical staff to focus on patient care.

- **Minimized Risk:** Robots can operate in environments hazardous to humans, such as isolation wards for contagious diseases.

- **Accessibility:** Robotic telemedicine and remote monitoring enable patients to access healthcare services regardless of their geographical location.

2. Key Applications of Healthcare Robotics:

- **Surgery:** Surgical robots assist surgeons in performing minimally invasive procedures with greater precision and smaller incisions, reducing patient trauma and recovery times.

- **Rehabilitation:** Robotic exoskeletons and assistive devices aid in physical therapy and rehabilitation, helping patients regain mobility and independence.

- **Medication Management:** Robots can dispense and administer medications with precise dosages and schedules, reducing medication errors.

- **Telemedicine:** Robotic telepresence systems enable remote consultations, allowing specialists to examine patients in remote or underserved areas.

- **Laboratory Automation:** Automated robotic systems can handle tasks such as sample processing, testing, and data analysis, improving efficiency in clinical laboratories.

3. Benefits of Healthcare Robotics:

- **Precision and Consistency:** Robots perform tasks with consistent precision, reducing the margin for error in diagnostics and treatment.

- **Minimized Infection Risk:** Robots can operate in sterile environments, minimizing the risk of healthcare-associated infections.

- **Less Invasive Procedures:** Minimally invasive robotic surgeries result in smaller incisions, less pain, and quicker recoveries for patients.

- **Remote Monitoring:** Telemedicine and remote monitoring technologies facilitate continuous care for patients, particularly those with chronic conditions.

- **Enhanced Training:** Medical professionals can use robotic simulators for training and skill development in a risk-free environment.

4. Challenges in Healthcare Robotics:

- **Cost:** The initial investment and maintenance of healthcare robotic systems can be expensive, limiting their accessibility.

- **Regulatory Compliance:** Ensuring that healthcare robots meet stringent regulatory standards for safety and effectiveness is a complex process.

- **Human-Machine Interaction:** Establishing seamless interaction between healthcare professionals and robots, particularly in surgery, requires specialized training and coordination.

- **Ethical Considerations:** Ethical concerns surround issues like patient privacy, informed consent for robotic procedures, and the potential for automation to replace human jobs.

5. Future Trends and Innovations:

The future of healthcare robotics holds promise in several areas, including the development of more affordable and accessible robotic systems, the integration of AI for diagnostic and predictive capabilities, and the expansion of robotic-assisted procedures to new medical fields.

In conclusion, healthcare robotics is poised to revolutionize patient care, diagnosis, and treatment, offering innovative solutions to address the challenges of the healthcare industry. As technology continues to advance and the healthcare community embraces these technologies, healthcare robotics will play an increasingly vital role in improving the overall quality of healthcare delivery.

C. Agricultural Robotics: Cultivating the Future of Farming

Agricultural robotics, often referred to as agri-robots or agbots, represents a transformative force in agriculture, offering innovative solutions to address the challenges of modern farming. These technologies leverage automation, artificial intelligence, and robotics to optimize crop production, enhance efficiency, and promote sustainable farming practices. In this in-depth exploration, we delve into the world of agricultural robotics, examining its significance, key applications, benefits, challenges, and the promising future it holds for the agricultural industry.

1. Significance of Agricultural Robotics:

- **Sustainable Farming:** Agbots enable precision agriculture, reducing resource wastage, minimizing environmental impact, and promoting sustainability.

- **Labor Shortages:** Automation in agriculture addresses the increasing shortage of farm labor, particularly for labor-intensive tasks like harvesting.

- **Crop Monitoring:** Robots equipped with sensors and imaging technologies provide real-time data on crop health, enabling proactive management.

- **Increased Productivity:** Agricultural robots can work around the clock, improving productivity and crop yields.

2. Key Applications of Agricultural Robotics:

- **Precision Farming:** Robots equipped with sensors and GPS technology can perform precise planting, irrigation, and fertilization, optimizing resource use.

- **Crop Monitoring:** Drones and ground-based robots capture data on crop health, identifying diseases, pests, and nutrient deficiencies.

- **Harvesting:** Robotic systems are designed for selective harvesting of fruits, vegetables, and other crops, reducing waste and labor costs.

- **Weed and Pest Control:** Autonomous robots and drones can identify and remove weeds and pests, reducing the need for chemical herbicides and pesticides.

- **Livestock Farming:** Robotics assist in the management of livestock, including automated feeding, monitoring, and even autonomous milking.

3. Benefits of Agricultural Robotics:

- **Resource Optimization:** Precision agriculture reduces water and fertilizer usage, minimizes soil compaction, and optimizes planting density.

- **Reduced Labor Costs:** Automation addresses labor shortages and reduces labor costs, particularly for repetitive and physically demanding tasks.

- **Data-Driven Decisions:** Data collected by agbots enable data-driven decision-making, enhancing crop management and yield predictions.

- **Sustainability:** Automation promotes sustainable farming practices by reducing chemical usage and environmental impact.

- **Global Food Security:** Increased efficiency in farming contributes to global food security by improving crop yields and reducing food waste.

4. Challenges in Agricultural Robotics:

- **High Initial Costs:** The adoption of agricultural robotics can be expensive, particularly for small-scale farmers.

- **Adaptability:** Agbots must be adaptable to different crops, climates, and farm configurations to be practical for widespread adoption.

- **Data Management:** Managing and analyzing the vast amount of data generated by agri-robots can be challenging for farmers.

- **Regulatory Hurdles:** Regulatory frameworks for agricultural robotics vary by region, posing challenges for manufacturers and farmers.

5. Future Trends and Innovations:

The future of agricultural robotics holds exciting possibilities, including the development of swarming robots that work in coordinated teams, the use of AI and machine learning for advanced crop analysis, and the integration of autonomous tractors and other machinery into farming operations.

In conclusion, agricultural robotics is poised to revolutionize farming, promoting sustainability, efficiency, and global food security. As technology continues to advance and agricultural communities embrace these innovations, agricultural robotics will play an increasingly crucial role in shaping the future of agriculture and addressing the challenges of feeding a growing global population.

D. Space Robotics: Exploring Beyond Earth's Boundaries

Space robotics represents a technological frontier in the exploration of outer space, enabling missions beyond the capabilities of human astronauts alone. These advanced robotic systems, equipped with specialized tools and artificial intelligence, have played critical roles in space exploration, satellite servicing, and planetary research. In this comprehensive exploration, we delve into the realm of space robotics, examining its significance, key applications, challenges, and the promising future it holds for the expansion of human knowledge beyond Earth's boundaries.

1. Significance of Space Robotics:

- **Risk Mitigation:** Space robots can perform tasks in environments where human presence is risky or impossible, reducing astronaut exposure to hazards.

- **Cost Efficiency:** Robotics offers cost-effective solutions for tasks that might otherwise require expensive human spaceflight missions.

- **Exploration:** Robots assist in the exploration of celestial bodies, such as Mars and the Moon, by conducting experiments and collecting samples.

- **Satellite Servicing:** Space robots can repair, refuel, and deorbit satellites, extending their operational lifetimes and reducing space debris.

2. Key Applications of Space Robotics:

- **Planetary Exploration:** Robotic rovers like NASA's Mars rovers (e.g., Curiosity and Perseverance) explore the surfaces of planets and moons, conducting experiments and collecting data.

- **Orbital Servicing:** Robots like the Canadarm on the Space Shuttle and the Dextre on the International Space Station (ISS) have performed complex tasks, including spacecraft assembly and maintenance.

- **Space Telescopes:** Robots have serviced and repaired space telescopes like the Hubble Space Telescope, extending their useful lifetimes.

- **Asteroid and Comet Exploration:** Space robots like the OSIRIS-REx mission collect samples from asteroids for analysis.

- **Satellite Debris Removal:** Emerging missions aim to capture and remove defunct satellites and debris from Earth's orbit.

3. Benefits of Space Robotics:

- **Extended Mission Lifetimes:** Robotics can repair, refuel, and maintain satellites and spacecraft, extending their operational capabilities.

- **Scientific Discovery:** Robotic explorers gather data and samples from celestial bodies, advancing our understanding of the solar system and beyond.

- **Risk Reduction:** Robots can perform high-risk tasks like inspecting and repairing spacecraft without risking human lives.

- **Resource Utilization:** Future missions may utilize robotic systems to mine and extract resources from asteroids or the Moon.

4. Challenges in Space Robotics:

- **Harsh Environments:** Space poses extreme challenges, including vacuum, radiation, and temperature extremes, which can affect robotic systems.

- **Communication Lag:** Remote control of robots in space may suffer from communication delays due to vast distances.

- **Autonomy:** Robots must exhibit a high level of autonomy to perform complex tasks with minimal human intervention.

- **Space Debris:** Maneuvering safely in increasingly cluttered orbits requires advanced navigation and collision avoidance capabilities.

5. Future Trends and Innovations:

The future of space robotics holds exciting developments, including autonomous swarming robots for planetary exploration, advanced AI for complex decision-making, and partnerships between humans and robots in deep space missions.

In conclusion, space robotics plays a vital role in advancing our knowledge of the cosmos, enabling safer and more cost-effective space exploration, and addressing the challenges of an increasingly crowded and debris-laden space environment. As technology continues to evolve, space robotics will remain a cornerstone of future missions into the vast expanse of outer space.

E. Service Robots: Enhancing Home, Retail, and Entertainment Experiences

Service robots have emerged as versatile assistants designed to improve our daily lives, augment customer service in retail environments, and enhance entertainment experiences. These robots, equipped with advanced sensors and artificial intelligence, perform a wide range of tasks, from household chores to providing information and entertainment in public spaces. In this comprehensive exploration, we delve into the world of service robots, examining their significance, key applications, benefits, challenges, and the promising future they hold for enhancing convenience and customer engagement.

1. Significance of Service Robots:

- **Convenience:** Service robots streamline everyday tasks, making life more comfortable and efficient for individuals and families.

- **Customer Experience:** In retail and entertainment, service robots enhance customer engagement and offer unique experiences.

- **Accessibility:** Service robots provide assistance to individuals with disabilities, promoting inclusivity and independence.

- **Labor Augmentation:** In sectors like retail and hospitality, robots complement human workers, freeing them from repetitive tasks to focus on more complex responsibilities.

2. Key Applications of Service Robots:

- **Home Robots:** These include vacuum and floor-cleaning robots, lawn-mowing robots, and personal assistant robots that manage household tasks and provide information.

- **Retail Robots:** Service robots in retail settings assist customers with wayfinding, product information, and inventory management, enhancing the shopping experience.

- **Entertainment Robots:** In entertainment venues, robots offer interactive experiences, from robotic bartenders to amusement park attractions.

- **Healthcare Robots:** Service robots in healthcare environments provide support with tasks such as patient care, medication delivery, and rehabilitation.

- **Hospitality Robots:** In hotels and restaurants, robots greet guests, deliver orders, and provide concierge services.

3. Benefits of Service Robots:

- **Efficiency:** Service robots complete tasks quickly and accurately, reducing the time and effort required for manual labor.

- **Customer Engagement:** Retail and entertainment robots offer interactive experiences that attract and engage customers.

- **Safety:** Robots can perform hazardous or repetitive tasks without risking human health or safety.

- **Accessibility:** Service robots provide assistance to people with mobility challenges, enabling greater independence.

- **Scalability:** In commercial applications, businesses can deploy multiple robots to meet increased demand without hiring additional staff.

4. Challenges in Service Robotics:

- **Cost:** The initial investment in service robots can be significant, limiting adoption by individuals and businesses.

- **Technology Limitations:** Service robots must be equipped with advanced sensors and AI to navigate complex environments and interact effectively with humans.

- **Ethical Concerns:** Concerns about job displacement, data privacy, and the ethical use of robots in sensitive settings need to be addressed.

- **User Acceptance:** People may be hesitant to interact with robots, particularly in service roles traditionally performed by humans.

5. Future Trends and Innovations:

The future of service robots includes increased personalization through AI, improved natural language processing, and enhanced human-robot collaboration. In retail, robots are likely to play larger roles in inventory management, and in entertainment, interactive and immersive experiences will continue to evolve.

In conclusion, service robots are transforming how we live, shop, and entertain ourselves. As technology continues to advance and society becomes more accustomed to interacting with robots, their presence is set to expand, offering increased convenience and engagement in various aspects of our lives.

F. Autonomous Vehicles and Drones: Transforming Transportation and Beyond

Autonomous vehicles (AVs) and drones represent groundbreaking innovations in the field of transportation and automation. These technologies, driven by artificial intelligence, sensors, and advanced control systems, have the potential to revolutionize industries ranging from logistics and agriculture to healthcare and urban planning. In this in-depth exploration, we delve into the world of autonomous vehicles and drones, examining their significance, key applications, benefits, challenges, and the promising future they hold for reshaping the way we move and interact with our environment.

1. Significance of Autonomous Vehicles and Drones:

- **Safety Improvement:** AVs have the potential to significantly reduce traffic accidents caused by human error, making roads safer for all.

- **Efficiency:** Autonomous transportation systems promise increased fuel efficiency and reduced traffic congestion, improving urban mobility.

- **Accessibility:** Drones and AVs can provide access to remote or otherwise inaccessible areas for various applications, from medical deliveries to disaster relief.

- **Environmental Impact:** AVs can be designed to be eco-friendly, reducing emissions and the environmental footprint of transportation.

2. Key Applications of Autonomous Vehicles and Drones:

- **Autonomous Vehicles:** AVs range from self-driving cars and trucks to autonomous buses and delivery vehicles used in logistics and ride-sharing services.

- **Drones:** Drones are employed in various industries, including agriculture for crop monitoring, healthcare for medical supply delivery, and filmmaking for aerial cinematography.

- **Public Transportation:** Autonomous buses and shuttles promise efficient and convenient urban transit solutions.

- **Logistics and Delivery:** AVs and drones are revolutionizing last-mile delivery, reducing delivery times and costs.

- **Surveillance and Security:** Drones are used for surveillance, border patrol, and emergency response, providing critical data and situational awareness.

3. Benefits of Autonomous Vehicles and Drones:

- **Safety:** AVs can significantly reduce accidents caused by human error, while drones offer safer alternatives for inspections in dangerous or inaccessible environments.

- **Efficiency:** Autonomous systems can optimize routes, reducing travel time and fuel consumption.

- **Accessibility:** Drones and AVs provide access to remote or underserved areas, opening up opportunities for medical services, disaster response, and more.

- **Environmental Impact:** AVs can be electric or hybrid, reducing emissions, and drones offer eco-friendly options for deliveries and surveillance.

- **Economic Growth:** These technologies have the potential to create new industries and jobs, from AV development to drone operations.

4. Challenges in Autonomous Vehicles and Drones:

- **Regulation and Liability:** Developing comprehensive regulations and determining liability in case of accidents are complex legal challenges.

- **Technical Hurdles:** Ensuring the safety and reliability of autonomous systems, particularly in complex urban environments, poses technical challenges.

- **Public Acceptance:** Convincing the public of the safety and benefits of AVs and drones remains a hurdle to widespread adoption.

- **Infrastructure:** Adapting infrastructure to accommodate AVs and drone operations, such as charging stations and landing zones, is a logistical challenge.

5. Future Trends and Innovations:

The future of AVs and drones includes increased connectivity, the emergence of flying cars and urban air mobility, and the development of autonomous vehicles for various specialized applications, such as mining and agriculture.

In conclusion, autonomous vehicles and drones have the potential to transform transportation, logistics, and various industries. As technology continues to advance and regulatory frameworks mature, these innovations will play an increasingly prominent role in reshaping how we move, work, and interact with our environment.

CHAPTER 11

Advanced Topics in Robotics and Future Trends in Automation: Pioneering the Next Frontier

As technology continues its relentless advance, the world of robotics and automation stands poised at the precipice of transformative innovation. In this exploration, we venture into the realm of advanced topics and future trends in robotics and automation, where cutting-edge research, emerging technologies, and visionary concepts converge. From soft robotics and biohybrids to the possibilities of quantum computing, this journey takes us beyond the boundaries of today's automation, offering a glimpse into the thrilling possibilities that lie ahead. Join us as we uncover the pioneering spirit and the ever-evolving landscape of robotics and automation.

A. Soft Robotics: The Future of Flexible Automation

Soft robotics is a rapidly advancing field that stands at the forefront of automation innovation. Unlike traditional rigid robots, soft robots are constructed from flexible and compliant materials, mimicking the dexterity and adaptability of natural organisms. This technology has the potential to revolutionize industries ranging from healthcare and manufacturing to search and rescue missions. In this comprehensive exploration, we delve into the world of soft robotics, examining its significance, key principles, applications, benefits, challenges, and the promising future it holds for creating more versatile and safer robots.

1. Significance of Soft Robotics:

- **Versatility:** Soft robots excel in environments where rigid robots struggle, such as unstructured surroundings or interactions with delicate objects.

- **Safety:** The compliance of soft robots reduces the risk of injury when they interact with humans or handle fragile materials.

- **Biological Inspiration:** Soft robotics draws inspiration from nature, offering the potential for more lifelike and biomimetic robots.

- **Accessibility:** These robots have applications in fields like healthcare, where gentleness and adaptability are critical.

2. Key Principles of Soft Robotics:

- **Compliant Materials:** Soft robots are typically made from flexible materials like elastomers, textiles, or hydrogels, allowing them to deform and conform to their surroundings.

- **Pneumatics and Hydraulics:** Soft robots often use air or liquid pressure to control their motion, enabling precise and adaptable movements.

- **Bio-Inspired Design:** Soft robots may take inspiration from biological organisms, replicating structures like tentacles, limbs, or appendages.

- **Sensing and Control:** Advanced sensing and control systems allow soft robots to adapt to changing conditions and perform complex tasks.

3. Applications of Soft Robotics:

- **Medical Robotics:** Soft robots are used in minimally invasive surgery, rehabilitation, and prosthetics, offering gentler interactions with biological tissues.

- **Search and Rescue:** Soft robots can navigate through debris and confined spaces, making them valuable for search and rescue missions.

- **Manufacturing:** These robots have applications in tasks like pick-and-place operations, where gentle handling of objects is required.

- **Environmental Monitoring:** Soft robotic sensors can be deployed in sensitive environments, such as marine ecosystems, for data collection.

4. Benefits of Soft Robotics:

- **Flexibility:** Soft robots can adapt to complex and unstructured environments, performing tasks that rigid robots cannot.

- **Safety:** Their compliant nature reduces the risk of damage to themselves, objects, or humans during interactions.

- **Biological Compatibility:** Soft robots are well-suited for medical applications, where they can safely interact with biological tissues.

- **Efficiency:** The flexibility of soft robots allows them to squeeze through tight spaces and navigate obstacles with ease.

5. Challenges in Soft Robotics:

- **Complex Control:** Controlling the motion and behavior of soft robots is a complex challenge, requiring advanced algorithms and sensors.

- **Material Durability:** Ensuring the durability and longevity of soft robot materials is an ongoing concern.

- **Scalability:** Mass production and scalability of soft robots can be challenging due to their intricate design and fabrication.

- **Integration:** Integrating soft robots with conventional robotics systems or electronics can be complex.

6. Future Trends and Innovations:

The future of soft robotics includes innovations in materials, such as self-healing polymers, and advanced control algorithms that enhance adaptability and autonomy. As the field evolves, soft robots are likely to find applications in a wide range of industries, from medical and manufacturing to space exploration.

In conclusion, soft robotics represents a paradigm shift in the world of automation, offering robots that are more adaptable, versatile, and safer for interacting with humans and the environment. As technology continues to advance and researchers push the boundaries of what soft robots can achieve, their significance in various fields is set to grow exponentially.

B. Biohybrid Robots: Bridging the Divide Between Nature and Technology

Biohybrid robots represent a cutting-edge fusion of biological organisms and artificial technology. These innovative machines are designed to incorporate living tissues, cells, or organisms into their structure, creating a new class of robots that can perform tasks with the precision of machinery and the adaptability of living beings. In this in-depth exploration, we delve into the world of biohybrid robots, examining their significance, key principles, applications, benefits, challenges, and the promising future they hold at the intersection of biology and robotics.

1. Significance of Biohybrid Robots:

- **Synergy of Biology and Technology:** Biohybrids harness the unique strengths of biological systems, such as self-healing and adaptability, while combining them with the control and precision of robotic technology.

- **Versatility:** These robots have the potential to excel in various applications, from healthcare and environmental monitoring to exploration and beyond.

- **Sustainability:** Biohybrids may offer more eco-friendly and sustainable alternatives to traditional robots by utilizing biological components.

- **Biomedical Advancements:** In medicine, biohybrid robots may offer groundbreaking solutions for drug delivery, tissue regeneration, and minimally invasive surgery.

2. Key Principles of Biohybrid Robots:

- **Biological Integration:** Biohybrids incorporate living cells, tissues, or organisms, which may include muscle cells, neurons, or microorganisms.

- **Biomechanical Interfaces:** These robots often feature interfaces that allow biological components to interact with synthetic materials or mechanical systems.

- **Control and Feedback:** Advanced control systems and sensors are essential for orchestrating the actions of the biological and synthetic elements.

- **Biochemical Signaling:** Some biohybrids use biochemical signaling to control biological functions, responding to specific cues or stimuli.

3. Applications of Biohybrid Robots:

- **Biomedical Engineering:** Biohybrids have potential applications in drug delivery, tissue engineering, and as platforms for studying biological processes.

- **Environmental Monitoring:** These robots can be used in aquatic environments to monitor water quality, study marine ecosystems, or perform environmental cleanup tasks.

- **Exploration:** Biohybrids may be employed in space exploration missions or extreme environments on Earth, where their adaptability and self-sustainability offer advantages.

- **Healthcare:** In the medical field, biohybrids may play roles in diagnostics, surgical assistance, and targeted drug delivery.

4. Benefits of Biohybrid Robots:

- **Adaptability:** Incorporating living components allows

biohybrids to respond to changing conditions and adapt to new challenges.

- **Biological Functionality:** These robots can perform tasks that are challenging for traditional robots, such as navigating complex environments or interacting with biological systems.

- **Sustainability:** Biohybrids may be designed to consume minimal resources and exhibit a reduced environmental footprint.

- **Biocompatibility:** In healthcare applications, biohybrids can interact with biological tissues without causing harm or rejection.

5. Challenges in Biohybrid Robotics:

- **Biological Control:** Coordinating the actions of biological components with mechanical systems is a complex control problem.

- **Ethical Considerations:** The use of living organisms in robots raises ethical questions related to animal welfare and the potential creation of new life forms.

- **Safety and Regulation:** Ensuring the safe operation of biohybrids and establishing regulatory frameworks is a challenge.

- **Integration Complexity:** Merging biological and synthetic components requires specialized expertise and may face technical hurdles.

6. Future Trends and Innovations:

The future of biohybrid robots includes advances in tissue engineering, improved control systems, and the development of biohybrids for specific applications, such as environmental cleanup or personalized medicine.

In conclusion, biohybrid robots represent a groundbreaking convergence of biology and technology, offering a new frontier for innovation. As this field continues to evolve, biohybrids hold the potential to revolutionize various industries and address complex challenges by harnessing the power of living organisms in tandem with artificial systems.

C. Humanoid Robots: Bridging the Gap Between Machines and Humans

Humanoid robots are a remarkable embodiment of engineering and artificial intelligence, designed to resemble and, to some extent, mimic human form and behavior. These robots have captivated our imagination for decades and hold immense potential in various fields, including healthcare, education, and entertainment. In this comprehensive exploration, we delve into the world of humanoid robots, examining their significance, key principles, applications, benefits, challenges, and the promising future they hold as companions, helpers, and even educators in our increasingly automated world.

1. Significance of Humanoid Robots:

- **Human Interaction:** Humanoids are designed to interact with humans naturally, making them suitable for roles requiring social engagement.

- **Versatility:** These robots can perform a wide range of tasks, from assisting in healthcare to providing customer service.

- **Empathy and Companionship:** Humanoids have the potential to provide companionship and emotional support to individuals, including the elderly and children.

- **Education and Research:** Humanoid robots are used as educational tools and for research in fields like psychology, neuroscience, and human-robot interaction.

2. Key Principles of Humanoid Robots:

- **Humanoid Design:** Humanoid robots are designed to resemble the human body to varying degrees, featuring a head, torso, arms, and legs.

- **Artificial Intelligence:** Advanced AI systems enable

humanoids to perceive their surroundings, understand speech, and make decisions.

- **Sensor Integration:** Humanoids are equipped with sensors, cameras, and microphones to perceive and interact with their environment.

- **Mechanical Actuation:** Motors and mechanisms replicate human movements, allowing humanoids to walk, gesture, and manipulate objects.

3. Applications of Humanoid Robots:

- **Healthcare:** Humanoids can assist with patient care, rehabilitation, and medication management, improving the quality of healthcare services.

- **Customer Service:** In retail and hospitality, humanoid robots can greet customers, provide information, and even take orders.

- **Education:** Humanoid robots are used as educational aids, helping children learn and adults acquire new skills.

- **Research:** Humanoids serve as research platforms for studying human behavior, cognition, and social interaction.

- **Entertainment:** Humanoid robots are used in theme parks, museums, and as performers in various entertainment events.

4. Benefits of Humanoid Robots:

- **Social Interaction:** Humanoids can engage with humans in ways that feel familiar and natural, fostering positive interactions.

- **Multifunctionality:** Their versatility allows humanoids to perform a wide range of tasks, making them valuable across industries.

- **Companion and Assistant:** Humanoid robots can provide companionship to the lonely and assistance to those with disabilities.

- **Education:** They can serve as interactive tutors, enhancing the learning experience.

5. Challenges in Humanoid Robotics:

- **Complexity:** The human form is complex, making the design and control of humanoid robots challenging.

- **Cost:** Developing and manufacturing humanoid robots can be expensive.

- **Ethical and Social Issues:** Humanoid robots raise ethical questions, particularly regarding privacy, job displacement, and the potential for emotional attachment.

- **Technical Hurdles:** Achieving natural movements, robust perception, and effective communication remains technically challenging.

6. Future Trends and Innovations:

The future of humanoid robots involves advancements in AI, natural language processing, and emotional understanding, enabling more sophisticated interactions. Humanoids may become commonplace in healthcare, caregiving, and educational settings, providing personalized support and companionship.

In conclusion, humanoid robots represent a bridge between technology and humanity, offering a glimpse into a future where robots can enhance our lives in myriad ways. As technology continues to advance, the integration of humanoids into various aspects of our society is likely to expand, shaping how we interact with and utilize these remarkable machines.

D. Robotics in Extreme Environments: Navigating the Depths and Depths of Space

Robotics has proven indispensable in exploring and conducting tasks in some of the most extreme and inhospitable environments on Earth and beyond. This in-depth exploration focuses on two key domains where robotic systems have made significant strides: underwater and space environments. These robots are designed to withstand harsh conditions, provide valuable data, and carry out missions that are too dangerous or impractical for humans. In this comprehensive examination, we delve into the significance, key principles, applications, benefits, challenges, and exciting future prospects of robotics in extreme environments.

1. Significance of Robotics in Extreme Environments:

- **Human Limitations:** Extreme environments, whether underwater or in space, present challenges such as extreme pressure, radiation, and temperature variations that are inhospitable to humans.

- **Scientific Exploration:** Robotic systems provide a means to study and collect data from remote and inaccessible regions, advancing our understanding of the Earth's oceans and celestial bodies.

- **Resource Exploration:** Robotics enable the search for valuable resources, such as minerals or water, in extraterrestrial environments.

- **Safety:** In dangerous environments, robots can perform tasks that would be life-threatening for humans, such as handling radioactive materials or exploring underwater caves.

2. Key Principles of Robotics in Extreme Environments:

- **Specialized Design:** Robots in extreme environments are engineered with materials and components that can withstand extreme pressures, temperatures, or radiation.

- **Autonomy:** These robots often require a high degree of autonomy to operate in environments with limited human intervention or communication.

- **Sensors:** Advanced sensors, often customized for the environment, enable robots to navigate and collect data.

- **Communication:** In remote environments, communication with the robot may be challenging, requiring specialized protocols or satellite links.

3. Applications of Robotics in Extreme Environments:

- **Underwater Robotics:** Submersibles and remotely operated vehicles (ROVs) are used for deep-sea exploration, marine archaeology, and offshore industry applications.

- **Space Robotics:** Robots, such as rovers and landers, are used for planetary exploration, asteroid mining, satellite maintenance, and more.

- **Environmental Monitoring:** In both underwater and space environments, robots gather data on climate, geology, and the health of ecosystems.

- **Resource Extraction:** Robots are being developed for mining resources on the Moon, Mars, and asteroids.

4. Benefits of Robotics in Extreme Environments:

- **Safety:** Robots protect human operators from dangerous conditions, such as high radiation or extreme pressure.

- **Data Collection:** These robots can collect valuable data that contributes to scientific research and resource discovery.

- **Efficiency:** In remote locations, robots can work around the clock, increasing efficiency and reducing the cost of operations.

- **Exploration:** They enable exploration and the potential discovery of new environments, species, or resources.

5. Challenges in Robotics in Extreme Environments:

- **Technical Complexity:** Developing robots that can function reliably in extreme conditions requires specialized engineering and materials.

- **Communication Lag:** In space missions, the delay in signal transmission to and from Earth can hinder real-time control.

- **Resource Limitations:** Operating in extreme environments often requires power sources and materials that are not readily available.

- **Environmental Variability:** Extreme environments can exhibit unpredictable and harsh conditions that challenge robot reliability.

6. Future Trends and Innovations:

The future of robotics in extreme environments involves innovations in AI for autonomous navigation, advanced sensors for data collection, and the development of versatile and adaptable robotic systems. These innovations will play crucial roles in upcoming lunar and Mars missions, as well as in expanding our understanding of Earth's oceans.

In conclusion, robotics in extreme environments is pushing the boundaries of exploration and knowledge. As technology continues to advance, these robots will open new frontiers, from uncovering the mysteries of our oceans to paving the way for future human missions to other planets. Their significance in both scientific and practical applications is set to grow, shaping our understanding of the universe and our own planet.

E. Quantum Computing and Robotics: Unleashing the Power of the Quantum Realm

Quantum computing is a revolutionary field of technology that promises to transform computation, enabling machines to solve problems that were once considered impossible for classical computers. When combined with robotics, quantum computing opens up new horizons for intelligent, autonomous machines that can tackle complex tasks and challenges with unprecedented efficiency and speed. In this comprehensive exploration, we delve into the intersection of quantum computing and robotics, examining its significance, key principles, applications, benefits, challenges, and the thrilling future it holds for automation and artificial intelligence.

1. Significance of Quantum Computing in Robotics:

- **Exponential Computing Power:** Quantum computers leverage the principles of quantum mechanics to perform certain computations exponentially faster than classical computers, enabling robots to process vast amounts of data quickly.

- **Optimization:** Quantum algorithms can solve optimization problems relevant to robotics, such as path planning, scheduling, and resource allocation, with remarkable efficiency.

- **Artificial Intelligence:** Quantum machine learning algorithms can enhance the learning capabilities of robots, making them more adaptable and intelligent.

- **Secure Communication:** Quantum encryption ensures secure communication for robots, safeguarding sensitive data in autonomous systems.

2. Key Principles of Quantum Computing and Robotics:

- **Quantum Bits (Qubits):** Unlike classical bits, which can be either 0 or 1, qubits can exist in a superposition of states, exponentially increasing computational possibilities.

- **Quantum Entanglement:** Qubits can be entangled, meaning the state of one qubit is dependent on the state of another, enabling faster communication and parallel processing.

- **Quantum Algorithms:** Specialized algorithms, like Grover's algorithm and Shor's algorithm, leverage quantum properties to solve specific problems efficiently.

- **Quantum Hardware:** Quantum computers require specialized hardware, such as superconducting qubits or trapped ions, to operate.

3. Applications of Quantum Computing in Robotics:

- **Optimization:** Quantum algorithms can optimize robotic operations, including route planning for autonomous vehicles and logistics in warehouses.

- **Machine Learning:** Quantum machine learning enhances robots' ability to analyze and adapt to complex data, such as image recognition and natural language processing.

- **Materials Discovery:** Quantum computing accelerates the discovery of new materials with properties ideal for robotics, such as lightweight and durable alloys.

- **Cryptography:** Quantum-resistant encryption safeguards robot communication, ensuring data integrity and security.

4. Benefits of Quantum Computing in Robotics:

- **Speed:** Quantum computing dramatically accelerates calculations, allowing robots to make real-time decisions in complex environments.

- **Efficiency:** Quantum algorithms optimize resource utilization, enhancing the energy efficiency of robotic systems.

- **Problem Solving:** Robots can tackle previously intractable problems, advancing fields like medicine, climate modeling, and materials science.

- **Security:** Quantum encryption protects robotic networks from cyber threats, ensuring data confidentiality and integrity.

5. Challenges in Quantum Computing and Robotics:

- **Technical Complexity:** Developing and maintaining quantum computing hardware is technically challenging and expensive.

- **Quantum Error Correction:** Quantum systems are

susceptible to errors, necessitating error correction techniques to maintain reliability.

- **Integration:** Integrating quantum computing technology into robotics platforms and programming languages is a complex endeavor.

- **Education and Expertise:** A shortage of experts in both quantum computing and robotics poses challenges in harnessing these technologies effectively.

6. Future Trends and Innovations:

The future of quantum computing and robotics includes advancements in quantum hardware, software development, and hybrid classical-quantum systems. As quantum computing technology matures, it will unlock new capabilities and possibilities for robots in various domains, from healthcare and manufacturing to space exploration.

In conclusion, the convergence of quantum computing and robotics represents a groundbreaking leap in automation and artificial intelligence. As these technologies continue to advance, robots will become more intelligent, efficient, and versatile, revolutionizing industries and addressing some of the world's most complex challenges. Quantum-enhanced robots hold the potential to reshape our future, offering solutions that were once beyond our reach.

F. Future Trends and Innovations in Automation: Shaping Tomorrow's World

Automation, a driving force behind increased productivity and efficiency, continues to evolve at an unprecedented pace. As we venture into the future, automation is poised to transform industries, redefine the nature of work, and enhance our daily lives. This comprehensive exploration delves into the most compelling future trends and innovations in automation, shedding light on the significance, key principles, potential applications, benefits, challenges, and the exciting prospects that await us in a world increasingly driven by intelligent machines.

1. Significance of Future Trends in Automation:

- **Economic Impact:** Automation trends will influence industries and economies, creating opportunities for growth and job transformation.

- **Technological Advancements:** Future trends will leverage cutting-edge technologies such as artificial intelligence, IoT, and quantum computing to redefine automation capabilities.

- **Social Implications:** Automation will have far-reaching social consequences, impacting workforce dynamics, education, and the nature of employment.

- **Global Challenges:** Automation is expected to play a vital role in addressing global challenges like climate change and healthcare.

2. Key Principles of Future Trends in Automation:

- **Artificial Intelligence:** Advanced AI algorithms will drive autonomous decision-making and problem-solving in automation systems.

- **Connectivity:** The Internet of Things (IoT) will connect devices and enable data sharing for smarter, interconnected automation systems.

- **Human-Machine Collaboration:** Automation trends will focus on enhancing human-machine collaboration and coexistence.

- **Sustainability:** Future automation solutions will prioritize sustainability, energy efficiency, and eco-friendly practices.

3. Potential Applications of Future Trends in Automation:

- **Smart Manufacturing:** Automation will revolutionize manufacturing with AI-driven predictive maintenance, quality control, and adaptive production.

- **Autonomous Transportation:** Self-driving vehicles and drones will transform transportation and logistics, improving safety and efficiency.

- **Healthcare:** Robotics and AI will enhance patient care, drug discovery, and telemedicine.

- **Agriculture:** Precision agriculture will leverage automation for sustainable farming practices and food security.

- **Smart Cities:** Automation will optimize city infrastructure, transportation, and utilities for more sustainable urban living.

4. Benefits of Future Trends in Automation:

- **Increased Efficiency:** Automation will streamline processes, reduce errors, and boost productivity across industries.

- **Improved Safety:** Automation will enhance workplace safety by taking on dangerous and repetitive tasks.

- **Enhanced Decision-Making:** AI-driven automation will provide valuable insights for better decision-making.

- **Resource Conservation:** Sustainability-focused automation will reduce resource consumption and environmental impact.

5. Challenges in Future Trends in Automation:

- **Job Displacement:** As automation advances, there will be concerns about job displacement and the need for workforce reskilling.

- **Data Security:** Interconnected automation systems will require robust cybersecurity measures to protect sensitive data.

- **Ethical Considerations:** Future automation trends will raise ethical questions regarding privacy, bias in AI, and the impact on society.

- **Regulatory Frameworks:** The development and deployment of advanced automation will necessitate updated regulations and standards.

6. Exciting Prospects and Innovations:

- **Quantum-Enhanced Automation:** Quantum computing and cryptography will bolster automation capabilities.

- **Human-Centered Automation:** Automation trends will prioritize human-centric design, focusing on user experience and safety.

- **Edge Computing:** Edge AI will enable faster, more localized decision-making in automation systems.

- **Resilience and Adaptability:** Automation will become more resilient, adaptable, and capable of handling unforeseen challenges.

In conclusion, the future of automation holds immense promise, with transformative trends and innovations poised to reshape industries, improve quality of life, and address pressing global challenges. Embracing these changes and adapting to the evolving automation landscape will be essential for individuals, organizations, and societies as we navigate the exciting future that awaits us in an increasingly automated world.

CHAPTER 12

Challenges, Case Studies, and Future Trends in Robotics and Automation: Navigating the Frontier of Innovation

The realm of robotics and automation is marked by relentless progress, where breakthroughs in technology are met with a myriad of challenges and the illumination of endless possibilities. In this introductory exploration, we embark on a journey through the intricacies of this dynamic field, where challenges are opportunities, case studies exemplify triumphs, and future trends beckon us toward uncharted territory. As we delve into the heart of robotics and automation, we shall uncover the challenges that researchers and practitioners face, illuminate the successes that have shaped industries, and chart a course toward the exciting future that awaits in this ever-evolving landscape.

A. Challenges in Robotics Research and Automation: Navigating the Frontiers of Innovation

Robotics research and automation are at the forefront of technological innovation, poised to revolutionize industries and transform our daily lives. However, this exciting field is not without its share of challenges. In this in-depth exploration, we delve into the multifaceted challenges that researchers and engineers encounter as they strive to advance robotics and automation, from technical hurdles to ethical considerations.

1. Technical Challenges:

- **Complex Control Systems:** Developing advanced control systems for robots that can operate autonomously in dynamic environments remains a formidable challenge. These systems must handle real-time data processing, adapt to unforeseen circumstances, and ensure safety.

- **Sensor Integration:** Integrating sensors for perception, localization, and mapping into robotic systems requires addressing issues related to sensor accuracy, noise, and compatibility. Fusion of data from multiple sensors further complicates the task.

- **Dexterity and Mobility:** Creating robots with human-like dexterity and mobility is challenging. Achieving fine manipulation, balance, and locomotion in unstructured environments is an ongoing pursuit.

- **Energy Efficiency:** Enhancing the energy efficiency of robots, especially those used in mobile applications, is crucial for prolonged operation and sustainability.

- **Autonomy and AI:** Advancing artificial intelligence (AI) for robotics, including decision-making and machine learning, is a continuous endeavor. Teaching robots to learn and adapt to novel situations is a complex problem.

2. Human-Robot Interaction:

- **Safety:** Ensuring the safety of humans working alongside robots is paramount. Collaborative robots, or cobots, must be equipped with safety mechanisms and standards to prevent accidents.

- **Ethical Considerations:** The integration of robots into society raises ethical concerns, including issues related to privacy, job displacement, and the potential for autonomous weapons.

3. Scalability and Adaptability:

- **Scalability:** Many robotic systems are designed for specific tasks or environments, making scalability a challenge. Creating adaptable robots that can operate in various contexts is essential.

- **Interoperability:** Ensuring that different robotic systems and platforms can communicate and work together seamlessly is crucial for widespread adoption.

4. Cost and Accessibility:

- **Cost:** Developing and manufacturing advanced robots can be expensive, limiting their accessibility to smaller companies and research institutions.

- **Skill Gap:** The shortage of skilled professionals in robotics and automation presents a barrier to progress. Addressing this skill gap is essential for the field's growth.

5. Regulatory and Legal Frameworks:

- **Regulations:** The lack of standardized regulations for robotics and automation can hinder their deployment and raise liability and safety concerns.

- **Intellectual Property:** Protecting intellectual property in the rapidly evolving field of robotics poses challenges, particularly in cases involving collaborative research and open-source development.

6. Environmental Impact:

- **Sustainability:** Minimizing the environmental footprint of robotic technologies, including their energy consumption and disposal of electronic waste, is becoming increasingly important.

7. Public Perception:

- **Acceptance:** Gaining public trust and acceptance of robots in various applications, including healthcare and autonomous vehicles, is crucial for their successful integration into society.

In conclusion, the challenges in robotics research and automation are diverse and complex, reflecting the rapid evolution of technology and its integration into our lives. Addressing these challenges requires collaboration among researchers, engineers, policymakers, and society at large. Overcoming these obstacles will pave the way for a future where robots and automation systems improve efficiency, enhance safety, and contribute to the betterment of society.

B. Real-world Implementations of Robotics and Automation: Transforming Industries and Society

The realm of robotics and automation, once confined to the realms of science fiction, has firmly entrenched itself in the fabric of our daily lives and industries. In this in-depth exploration, we journey through the myriad real-world implementations of robotics and automation that are revolutionizing sectors ranging from manufacturing and healthcare to agriculture and transportation. These implementations, driven by technological advancements and innovation, underscore the significance, benefits, challenges, and promising future of automation in diverse fields.

1. Manufacturing and Industry:

- **Industrial Robots:** Manufacturing has witnessed a profound transformation with the widespread adoption of industrial robots. These machines handle tasks like welding, assembly, and material handling with precision and efficiency, reducing costs and improving product quality.

- **Industry 4.0:** The fourth industrial revolution, known as Industry 4.0, integrates automation, data exchange, and artificial intelligence to create smart factories. These factories optimize production through real-time data analysis and predictive maintenance.

- **3D Printing:** Additive manufacturing, or 3D printing, automates the creation of intricate parts and prototypes, offering flexibility and customization while reducing waste.

2. Healthcare:

- **Robotic Surgery:** Surgeons are using robotic-assisted systems for minimally invasive procedures, enhancing precision and reducing patient recovery times.

- **Rehabilitation Robotics:** Robots aid in physical therapy and rehabilitation, helping patients recover from injuries and improve mobility.

- **Pharmaceuticals:** Automation streamlines drug discovery, production, and quality control processes in the pharmaceutical industry.

3. Agriculture:

- **Precision Agriculture:** Automated machinery equipped with sensors and GPS technology optimizes crop planting, monitoring, and harvesting, increasing yields and resource efficiency.

- **Drones:** Agricultural drones conduct aerial surveys, crop analysis, and pesticide application with precision, reducing manual labor and environmental impact.

4. Transportation:

- **Autonomous Vehicles:** Self-driving cars and trucks promise safer and more efficient transportation systems, with potential applications in ride-sharing and logistics.

- **Drone Delivery:** Companies are exploring automated drone delivery services for fast and efficient transportation of goods.

5. Logistics and Warehousing:

- **Robotic Pickers:** Automated robots equipped with computer vision and machine learning capabilities assist in picking, packing, and sorting products in warehouses, reducing human labor and errors.

- **Last-Mile Delivery Robots:** Small robots navigate sidewalks and deliver packages to customers' doorsteps, optimizing the last leg of delivery operations.

6. Retail and Customer Service:

- **Automated Checkout:** Self-checkout kiosks and smart shopping carts streamline the retail experience, reducing wait times and improving customer satisfaction.

- **Chatbots and Virtual Assistants:** AI-driven chatbots and virtual assistants provide customer support and information on websites and mobile apps.

7. Energy and Environment:

- **Renewable Energy:** Robots maintain and repair solar panels and wind turbines, increasing the efficiency and lifespan of renewable energy installations.

- **Environmental Monitoring:** Autonomous robots and drones collect data for environmental monitoring and disaster response.

8. Space Exploration:

- **Rovers:** Robotic rovers like NASA's Curiosity explore the

Martian surface, conducting experiments and collecting data.

*   **Satellite Servicing:** Autonomous robots in space perform maintenance and repair tasks on satellites, extending their operational lifetimes.

In conclusion, real-world implementations of robotics and automation are reshaping industries and society in profound ways. These innovations are enhancing productivity, improving safety, and addressing complex challenges. As technology continues to advance, the potential applications of automation in various domains are boundless, promising a future where intelligent machines play an increasingly vital role in our lives.

C. Industry-specific Case Studies: Transforming Sectors through Robotics and Automation

The impact of robotics and automation is felt across a spectrum of industries, each presenting unique challenges and opportunities. In this in-depth exploration, we delve into industry-specific case studies that illustrate how these technologies are revolutionizing sectors as diverse as manufacturing, healthcare, agriculture, and more. These real-world examples demonstrate the transformative power of automation in addressing industry-specific challenges and driving innovation.

1. Manufacturing:

Case Study 1: Automotive Manufacturing

In the automotive industry, robots have become indispensable. Automotive manufacturers like Tesla have embraced automation to streamline production. Tesla's Gigafactories employ thousands of robots for tasks such as welding, painting, and assembly. This automation has not only increased production efficiency but also improved precision and product quality.

2. Healthcare:

Case Study 2: Robotic Surgery in Healthcare

Robotic surgery systems, such as the da Vinci Surgical System, have transformed healthcare. Surgeons use these robots to perform minimally invasive procedures with remarkable precision. For example, in prostate surgery, the da Vinci system has led to reduced blood loss, shorter hospital stays, and quicker recovery times for patients.

3. Agriculture:

Case Study 3: Precision Agriculture

John Deere, a leader in agricultural machinery, has integrated automation into its equipment to enhance farming practices. Their precision agriculture solutions leverage GPS technology, sensors, and automation to optimize planting, harvesting, and crop management. Farmers benefit from increased yields, reduced resource usage, and improved sustainability.

4. Logistics and Warehousing:

Case Study 4: Amazon's Warehouse Automation

Amazon's vast network of fulfillment centers relies heavily on automation. Robots, like the Kiva system, autonomously transport products to human workers, reducing the time required to pick and pack customer orders. This automation enables Amazon to fulfill orders more efficiently, especially during peak shopping seasons.

5. Energy:

Case Study 5: Wind Turbine Maintenance

The wind energy sector faces challenges in maintaining offshore wind turbines. Companies like GE Renewable Energy are using robotic systems to inspect and repair these turbines. Robots equipped with cameras and sensors can climb turbine towers, assess the condition of blades, and perform repairs, reducing the need for costly human interventions.

6. Retail:

Case Study 6: Walmart's Shelf-Scanning Robots

Walmart employs autonomous robots equipped with cameras

and sensors to scan store shelves for inventory management and restocking purposes. These robots efficiently navigate store aisles, identify out-of-stock items, and optimize inventory levels. This automation helps Walmart reduce stockouts and improve the shopping experience for customers.

7. Space Exploration:

Case Study 7: Mars Rover Missions

NASA's Mars rovers, including Curiosity and Perseverance, are prime examples of automation in space exploration. These robots conduct scientific experiments, capture images, and traverse the Martian terrain to gather data. The success of these missions demonstrates the feasibility and benefits of robotic exploration in extreme environments.

8. Food Industry:

Case Study 8: Food Processing Automation

Food processing plants, such as those in the meat and poultry industry, rely on automation for tasks like cutting, sorting, and packaging. Companies like Tyson Foods use robots to automate these processes, improving food safety, quality, and production efficiency.

These industry-specific case studies showcase the diverse ways in which robotics and automation are transforming sectors worldwide. Whether it's enhancing manufacturing efficiency, improving patient outcomes in healthcare, or revolutionizing agriculture, automation continues to shape industries and drive innovation, making the promise of a more automated future increasingly tangible.

D. Challenges Faced and Overcome in Robotics and Automation

The journey of robotics and automation, marked by remarkable advancements and innovations, is not devoid of challenges. Engineers, researchers, and organizations have encountered multifaceted obstacles on their path to harnessing the full potential of automation. In this comprehensive exploration, we delve into the challenges faced and the ingenious solutions that have been devised to overcome them, showcasing the resilience and creativity of the robotics and automation community.

1. Technical Challenges:

Challenge 1: Complex Control Systems

Overcoming the Challenge: Engineers have developed advanced control algorithms and machine learning techniques to enable robots to handle complex tasks autonomously. Real-time data processing and adaptive control systems have improved robot performance in dynamic environments.

Challenge 2: Sensor Integration

Overcoming the Challenge: Sensor technology has evolved, with more accurate and reliable sensors available. Additionally, sensor fusion techniques combine data from various sensors, enhancing perception and localization capabilities in robots.

Challenge 3: Dexterity and Mobility

Overcoming the Challenge: Advances in robotics have led to the development of highly dexterous robotic arms and mobile platforms that can navigate unstructured environments. These robots use advanced kinematics and dynamics to achieve human-like movements.

2. Human-Robot Interaction:

Challenge 4: Safety in Collaborative Robotics

Overcoming the Challenge: Collaborative robots, or cobots, are equipped with safety features like force sensors and collision detection. These mechanisms ensure that robots can work alongside humans safely. Standardized safety protocols and risk assessments have also been established.

Challenge 5: Ethical Considerations

Overcoming the Challenge: Ethical guidelines and regulations are being developed to address privacy concerns, job displacement, and ethical use of robots. Roboticists and ethicists collaborate to ensure responsible AI and robotics development.

3. Scalability and Adaptability:

Challenge 6: Scalability of Robotic Systems

Overcoming the Challenge: Modular and reconfigurable robot designs have emerged, enabling scalability across different applications. Universal robot platforms and software-defined robotics facilitate adaptation to diverse tasks.

Challenge 7: Interoperability*

Overcoming the Challenge: Standardized communication protocols and interfaces are being established to ensure interoperability between different robotic systems. These standards facilitate the seamless integration of robots into existing workflows.

4. Cost and Accessibility:

Challenge 8: High Development and Maintenance Costs

Overcoming the Challenge: Economies of scale and advancements in manufacturing have lowered the cost of robotic components and systems. Open-source software and hardware initiatives, such as Robot Operating System (ROS), have democratized access to robotics development.

Challenge 9: Skill Gap*

Overcoming the Challenge: Educational programs and training initiatives have been launched to bridge the skill gap in robotics and automation. Online courses, robotics competitions, and vocational training programs empower individuals to enter the field.

5. Regulatory and Legal Frameworks:

Challenge 10: Lack of Standard Regulations

Overcoming the Challenge: Governments and international organizations are developing regulations and standards for robotics and automation. These frameworks ensure safety, security, and ethical use of robotic technologies.

Challenge 11: Intellectual Property*

Overcoming the Challenge: Intellectual property laws and patent systems protect innovations in robotics. Collaboration and licensing agreements facilitate technology sharing while safeguarding intellectual property rights.

6. Environmental Impact:

Challenge 12: Energy Efficiency and Sustainability

Overcoming the Challenge: Research and development efforts focus on improving the energy efficiency of robotic systems. Sustainable materials and manufacturing processes are employed to reduce the environmental footprint of robotics.

7. Public Perception:

Challenge 13: Building Trust in Robots

Overcoming the Challenge: Transparency in AI and robotics, along with public education and awareness campaigns, helps build trust in robotic technologies. Demonstrating the tangible benefits of robots in various applications fosters acceptance.

In conclusion, the challenges faced and overcome in robotics and automation underscore the resilience and innovation of the field. Each obstacle has spurred creative solutions and advancements, leading to the continued growth and impact of automation in various industries and society at large. As technology continues to evolve, robotics and automation will persist in shaping a more automated, efficient, and connected future.

E. Innovations and Lessons Learned in Robotics and Automation

The realms of robotics and automation have witnessed remarkable innovations that continue to redefine industries and shape the way we live and work. Alongside these innovations, valuable lessons have been learned, often through challenges and setbacks. In this comprehensive exploration, we delve into the most influential innovations and the key lessons derived from them, illuminating the evolution and maturation of robotics and automation.

1. Innovations in Robotics and Automation:

Innovation 1: Collaborative Robotics (Cobots)

Impact: Collaborative robots, or cobots, are designed to work safely alongside humans. They have transformed industries by augmenting human capabilities and performing tasks that were previously considered too dangerous or monotonous. Cobots have been deployed in manufacturing, healthcare, and logistics, revolutionizing production processes and improving workplace safety.

Innovation 2: Machine Learning and AI Integration

Impact: The integration of machine learning and artificial intelligence (AI) has empowered robots with the ability to learn, adapt, and make autonomous decisions. AI-driven robots excel in tasks such as image recognition, natural language processing, and complex problem-solving. This innovation has expanded the range of applications for robotics, including autonomous vehicles, smart assistants, and predictive maintenance.

Innovation 3: Advanced Sensors and Perception Technologies

Impact: The development of advanced sensors, including LiDAR, depth cameras, and high-resolution imaging systems, has enhanced robots' perception capabilities. These sensors enable robots to navigate complex environments, detect objects with precision, and interact with their surroundings. Innovations in sensor technology have been pivotal in enabling autonomous vehicles, drones, and robotic vision systems.

Innovation 4: Human-Robot Interaction (HRI) Technologies

Impact: HRI technologies, such as speech recognition, gesture control, and intuitive interfaces, have facilitated seamless communication between humans and robots. This innovation has found applications in home automation, healthcare, and customer service, enhancing user experiences and accessibility.

Innovation 5: Quantum Computing in Robotics

Impact: Quantum computing is poised to revolutionize robotics by addressing complex optimization problems that were previously intractable. Quantum algorithms can expedite tasks like path planning, drug discovery, and materials science, ushering in a new era of computational efficiency for robots.

2. Lessons Learned:

Lesson 1: Safety is Paramount

Lesson Learned: The development of safety standards and mechanisms is crucial when integrating robots into collaborative and shared workspaces. Safety features, such as force sensors and collision detection, have become essential to prevent accidents and ensure the well-being of humans working alongside robots.

Lesson 2: Ethical Considerations Matter

Lesson Learned: Ethical considerations, including privacy, job displacement, and the responsible use of AI, have gained prominence. The robotics and automation community has learned the importance of addressing these ethical concerns through transparent development practices, regulations, and public dialogue.

Lesson 3: Interdisciplinary Collaboration is Key

Lesson Learned: Robotics and automation require collaboration across various disciplines, including engineering, computer science, ethics, and policy-making. Interdisciplinary cooperation fosters holistic solutions, ensuring that robots are not only technologically advanced but also socially responsible.

Lesson 4: Adaptability and Scalability are Essential

Lesson Learned: The ability to adapt robotic systems for diverse applications and scale them according to specific needs is crucial. Modular designs and open-source platforms have emerged as effective strategies for achieving adaptability and scalability.

Lesson 5: Continuous Learning and Adaptation

Lesson Learned: The field of robotics and automation is dynamic, with rapid technological advancements. The lesson learned is the importance of continuous learning and adaptation to stay at the forefront of innovation, whether through professional development or educational initiatives.

In conclusion, innovations in robotics and automation have reshaped industries and everyday life, while the lessons learned emphasize the significance of safety, ethics, collaboration, adaptability, and ongoing education. As robotics and automation continue to evolve, these insights guide the field toward a future where intelligent machines enhance our capabilities, address complex challenges, and coexist harmoniously with humanity.

Conclusion

As we draw the final curtain on this journey through these pages, we invite you to reflect on the knowledge, insights, and discoveries that have unfolded before you. Our exploration of various subjects has been a captivating voyage into the depths of understanding.

In these chapters, we have ventured through the intricacies of numerous topics and examined the key concepts and findings that define these fields. It is our hope that you have found inspiration, enlightenment, and valuable takeaways that resonate with you on your own quest for knowledge.

Remember that the pursuit of understanding is an ever-evolving journey, and this book is but a milestone along the way. The world of knowledge is vast and boundless, offering endless opportunities for exploration and growth.

As you conclude this book, we encourage you to carry forward the torch of curiosity and continue your exploration of these subjects. Seek out new perspectives, engage in meaningful discussions, and embrace the thrill of lifelong learning.

We express our sincere gratitude for joining us on this intellectual adventure. Your curiosity and dedication to expanding your horizons are the driving forces behind our shared quest for wisdom and insight.

Thank you for entrusting us with a portion of your intellectual journey. May your pursuit of knowledge lead you to new heights and inspire others to embark on their own quests for understanding.

With sincere appreciation,

Nikhilesh Mishra, Author

Recap of Key Takeaways

Throughout our journey into the realm of robotics and automation, we've explored a vast landscape of topics, technologies, and applications. As we conclude this comprehensive guide, let's recap the key takeaways that encapsulate the essence of these fields and their significance in today's world.

1. Definition and Scope of Robotics:

- Robotics is the interdisciplinary field that combines mechanical engineering, electronics, computer science, and artificial intelligence to create machines capable of performing tasks autonomously or semi-autonomously.

- The scope of robotics extends from industrial robots used in manufacturing to service robots in healthcare, agriculture, and beyond.

2. Historical Perspective of Robotics and Automation:

- The history of robotics dates back centuries, with early

automata and mechanical inventions serving as precursors to modern robots.

- The industrial revolution brought about significant advancements in automation, and subsequent decades witnessed the development of programmable robotic arms.

3. Key Concepts (Sensors, Actuators, Control Systems):

- Sensors gather data from the environment, actuators execute physical actions, and control systems orchestrate the entire process.

- Robotics and automation heavily rely on sensors like cameras, LiDAR, and IMUs, which enable perception and navigation.

- Actuators, including electric motors and pneumatic systems, provide the mechanical motion required for tasks.

4. Robotics and Automation Applications:

- Robotics and automation are applied across a wide range of industries, including manufacturing, healthcare, agriculture, transportation, and space exploration.

- These technologies enhance efficiency, precision, and safety in various tasks, from robotic surgery in healthcare to autonomous vehicles on the roads.

5. Benefits and Challenges of Robotics and Automation:

- Benefits include increased productivity, improved quality, enhanced safety, and the ability to perform tasks in hazardous environments.

- Challenges encompass technical complexities, safety concerns, ethical considerations, job displacement, and the need for interdisciplinary collaboration.

6. Robotics Fundamentals:

- Robotics encompasses various types of robots, including industrial robots used in manufacturing, service robots in healthcare, and autonomous robots in research and exploration.

- Robot components, including manipulators, end effectors, and drives, determine a robot's capabilities and functions.

7. Sensors, Perception, and Automation Technologies:

- Sensors enable robots to perceive the environment, detect objects, and navigate.

- Perception technologies, such as computer vision and LiDAR, enhance robots' understanding of their surroundings.

- Automation technologies, including sensors, actuators, and control systems, form the backbone of robotic systems.

8. Robot Manipulation, Mobility, and Process Automation:

- Robot manipulation involves tasks like grasping and manipulating objects with precision.

- Mobility solutions range from wheeled robots to legged robots capable of traversing challenging terrains.

- Process automation streamlines industrial workflows, reducing manual labor and errors.

9. Robot Learning, AI, and Artificial Intelligence in Automation:

- Machine learning and AI enable robots to learn from data and adapt to changing environments.

- Reinforcement learning is used to train robots for specific tasks, such as autonomous navigation.

- Ethical considerations, including human-robot interaction and AI ethics, are crucial in the development of intelligent automation.

10. Robot Vision, Perception, and Cognitive Automation:

- 3D perception, object recognition, and visual SLAM enhance robots' vision capabilities.

- Cognitive automation leverages AI to enable robots to make intelligent decisions based on visual data.

11. Robot Control, Programming, and Business Process Automation (BPA):

- Robot programming languages like ROS, Python, and C++ are used to command and control robots.

- Real-time control ensures robots respond quickly to changing conditions.

- Business process automation streamlines organizational workflows using robots and software.

12. Robot Collaboration, Swarm Robotics, and Home Automation:

- Multi-robot systems collaborate to perform tasks efficiently.

- Swarm robotics mimics collective behavior observed in nature to solve complex problems.

- Home automation systems use robots and IoT devices to enhance convenience and security.

13. Robot Ethics, Safety, and Regulations in Automation:

- Robot safety considerations are paramount to prevent accidents in collaborative environments.

- Ethical issues, including privacy and job displacement, require careful consideration.

- Regulations and standards, such as ISO and ANSI, ensure safe and responsible robotics.

14. Robotics and Automation Applications:

- Industrial robotics automate manufacturing processes, increasing efficiency and precision.

- Healthcare robotics assist in surgery, therapy, and medical procedures.

- Agricultural robotics optimize farming practices for higher yields and sustainability.

- Space robotics explore celestial bodies and maintain space infrastructure.

- Service robots enhance daily life, from home and retail to entertainment.

- Autonomous vehicles and drones revolutionize transportation and logistics.

15. Advanced Topics in Robotics and Future Trends in Automation:

- Emerging areas like soft robotics, biohybrid robots, and humanoid robots offer novel capabilities.

- Robotics in extreme environments, quantum computing, and future trends shape the future of automation.

16. Challenges, Case Studies, and Future Trends in Robotics and Automation:

- Challenges in research and implementation require innovative solutions and interdisciplinary collaboration.

- Real-world case studies illustrate the impact of automation in diverse industries.

- Future trends promise continued innovation and transformation in robotics and automation.

As we reflect on these key takeaways, it becomes clear that robotics and automation are dynamic, interdisciplinary fields with immense potential for improving our lives, reshaping industries, and advancing technology. The journey into the world of robotics and automation continues, offering new horizons and opportunities for those who seek to explore, innovate, and shape the future..

The Future of Robotics and Automation

The future of robotics and automation promises a profound transformation in how we live, work, and interact with the world. As we look ahead, it becomes evident that these fields are poised to redefine industries, enhance our daily lives, and tackle complex challenges. In this exploration of the future, we delve into key trends and developments that will shape the path of robotics and automation in the years to come.

1. Advanced Autonomy and AI Integration:

Future Trend 1: Increased Autonomy

Robots will become increasingly autonomous, capable of performing complex tasks with minimal human intervention. This trend is driven by advancements in AI, machine learning, and sensor technologies. Autonomous vehicles, for instance, will navigate city streets and highways, revolutionizing transportation and logistics.

Future Trend 2: AI-Driven Decision-Making

AI will play a central role in decision-making for robots and automated systems. These systems will adapt to dynamic environments, learn from data, and make real-time decisions. This AI-driven decision-making will enhance robots' ability to perform tasks in unpredictable scenarios, such as disaster response and exploration.

2. Human-Robot Collaboration and Coexistence:

Future Trend 3: Collaborative Robots (Cobots)

Collaborative robots (cobots) will become more prevalent in workplaces, working alongside humans in shared spaces. Enhanced safety features and natural language interfaces will facilitate seamless collaboration. Cobots will take on repetitive and physically demanding tasks, freeing up human workers for more creative and strategic roles.

Future Trend 4: Social and Companion Robots

The development of social and companion robots will cater to various societal needs. These robots will assist the elderly and individuals with disabilities, provide companionship, and support mental health initiatives. Human-robot interaction will evolve to encompass emotional understanding and empathy.

3. Biologically-Inspired and Soft Robotics:

Future Trend 5: Soft Robotics

Soft robotics, inspired by nature, will gain prominence. These robots will be more adaptable to unstructured environments and interact safely with humans. Soft robotic applications include medical devices, search and rescue operations, and exploration of delicate environments.

Future Trend 6: Biohybrid Robots

Biohybrid robots will merge biological and synthetic components, blurring the line between living organisms and machines. These robots will have potential applications in healthcare, environmental monitoring, and scientific research.

4. Quantum Computing and Robotics:

Future Trend 7: Quantum Computing for Optimization

Quantum computing will address complex optimization problems, enabling robots to plan and execute tasks more efficiently. This technology will revolutionize logistics, resource allocation, and supply chain management.

5. Environmental Sustainability:

Future Trend 8: Sustainable Robotics

The development of environmentally sustainable robotics will reduce the ecological footprint of automation. These robots will use eco-friendly materials, energy-efficient systems, and recycling mechanisms. Agricultural robots, for example, will contribute to precision farming and reduced resource usage.

6. Ethics, Regulation, and Policy:

Future Trend 9: Ethical AI and Robotics

Ethical considerations will continue to be a critical focus.

Regulations and standards will evolve to ensure the responsible development and deployment of AI and robotics. Organizations and governments will collaborate to address privacy, accountability, and ethical use.

7. Education and Workforce Development:

Future Trend 10: Education for Automation

Education and workforce development programs will adapt to prepare individuals for the changing job landscape. Robotics and automation education will become more accessible, addressing the need for skilled professionals in these fields.

In conclusion, the future of robotics and automation holds the promise of a more connected, efficient, and adaptable world. These technologies will not only reshape industries but also empower individuals and societies to address complex challenges and enhance our quality of life. As we continue to innovate and explore the vast potential of robotics and automation, it is clear that the future is bright, offering boundless opportunities for those who embrace and shape it.

Glossary of Terms

The fields of robotics and automation are rich with technical terminology that encompasses a wide range of concepts, technologies, and applications. This glossary provides a comprehensive reference guide to key terms, helping to demystify the language of robotics and automation.

1. **AI (Artificial Intelligence):** AI refers to the simulation of human intelligence in machines, enabling them to learn, reason, and make decisions. In robotics, AI plays a crucial role in enabling robots to perceive and interact with their environment intelligently.

2. **Actuator:** An actuator is a device that converts an input signal, typically electrical or pneumatic, into mechanical motion. Actuators are responsible for the physical movement of robots and automated systems.

3. **Automation:** Automation is the process of using technology, typically robotics and control systems, to perform tasks with minimal human intervention. It aims to improve efficiency, precision, and consistency.

4. **Cobot (Collaborative Robot):** A cobot is a robot designed to work alongside humans safely. These robots often have built-in safety features and sensors to prevent collisions with humans.

5. Control System: A control system is a set of components that manage and regulate the behavior of a robot or automated system. It receives input from sensors and issues commands to actuators, ensuring the desired outcome.

6. **End Effector:** The end effector is the tool or attachment at the end of a robot's arm or manipulator. It is designed for specific tasks, such as gripping objects, welding, or performing precise actions.

7. **Human-Robot Interaction (HRI):** HRI refers to the study and development of interfaces and communication methods that allow humans to interact with robots effectively and intuitively.

8. **IoT (Internet of Things):** IoT is a network of interconnected devices and sensors that can communicate and share data over the internet. In automation, IoT plays a role in connecting and controlling various devices and systems.

9. **Machine Learning:** Machine learning is a subset of AI that involves training algorithms to learn from data and make predictions or decisions. It is used in robotics for tasks like object recognition and path planning.

10. **ROS (Robot Operating System):** ROS is an open-source framework used in robotics for developing and controlling robotic systems. It provides a set of tools and libraries for building and managing robots.

11. **Sensor:** A sensor is a device that detects and measures physical properties or environmental conditions, such as temperature, light, or proximity. Sensors provide data that robots use for perception and decision-making.

12. **SLAM (Simultaneous Localization and Mapping):** SLAM is a technique used in robotics to create maps of an environment while simultaneously determining the robot's position within that environment. It is crucial for autonomous navigation.

13. **Soft Robotics:** Soft robotics involves the development of robots with flexible and deformable structures, often inspired by natural organisms. Soft robots are suited for tasks in unstructured environments and for interactions with humans.

14. **Sustainability:** Sustainability in robotics and automation focuses on developing technologies and practices that reduce environmental impact, such as using energy-efficient components and materials.

15. **Swarm Robotics:** Swarm robotics is the study of coordinating large groups of relatively simple robots to work together to accomplish tasks. It draws inspiration from collective behavior observed in nature.

16. **Teleoperation:** Teleoperation is the remote control of a robot or automated system by a human operator. It is often used in

scenarios where direct human involvement is necessary, such as hazardous environments or space exploration.

17. **UVS (Unmanned Vehicle Systems):** UVS refers to autonomous or remotely operated vehicles used in various applications, including aerial drones, autonomous cars, and underwater exploration.

This glossary provides a foundation for understanding the terminology used in the fields of robotics and automation. As these fields continue to evolve, new terms and concepts will emerge, highlighting the dynamic nature of technology and innovation.

Resources and References

As you reach the final pages of this book by Nikhilesh Mishra, consider it not an ending but a stepping stone. The pursuit of knowledge is an unending journey, and the world of information is boundless.

Discover a World Beyond These Pages

We extend a warm invitation to explore a realm of boundless learning and discovery through our dedicated online platform: **www.nikhileshmishra.com**. Here, you will unearth a carefully curated trove of resources and references to empower your quest for wisdom.

Unleash the Potential of Your Mind

- **Digital Libraries:** Immerse yourself in vast digital libraries, granting access to books, research papers, and academic treasures.

- **Interactive Courses:** Engage with interactive courses and lectures from world-renowned institutions, nurturing your thirst for knowledge.

- **Enlightening Talks:** Be captivated by enlightening talks delivered by visionaries and experts from diverse fields.

- **Community Connections:** Connect with a global community of like-minded seekers, engage in meaningful discussions, and share your knowledge journey.

Your Journey Has Just Begun

Your journey as a seeker of knowledge need not end here. Our website awaits your exploration, offering a gateway to an infinite universe of insights and references tailored to ignite your intellectual curiosity.

Acknowledgments

As I stand at this pivotal juncture, reflecting upon the completion of this monumental work, I am overwhelmed with profound gratitude for the exceptional individuals who have been instrumental in shaping this remarkable journey.

In Loving Memory

To my father, **Late Shri Krishna Gopal Mishra,** whose legacy of wisdom and strength continues to illuminate my path, even in his physical absence, I offer my deepest respect and heartfelt appreciation.

The Pillars of Support

My mother**, Mrs. Vijay Kanti Mishra,** embodies unwavering resilience and grace. Your steadfast support and unwavering faith in my pursuits have been the bedrock of my journey.

To my beloved wife, **Mrs. Anshika Mishra,** your unshakable belief in my abilities has been an eternal wellspring of motivation. Your constant encouragement has propelled me to reach new heights.

My daughter, **Miss Aarvi Mishra,** infuses my life with boundless joy and unbridled inspiration. Your insatiable curiosity serves as a constant reminder of the limitless power of exploration and discovery.

Brothers in Arms

To my younger brothers, **Mr. Ashutosh Mishra** and **Mr. Devashish Mishra,** who have steadfastly stood by my side, offering unwavering support and shared experiences that underscore the strength of familial bonds.

A Journey Shared

This book is a testament to the countless hours of dedication and effort that have gone into its creation. I am immensely grateful for the privilege of sharing my knowledge and insights with a global audience.

Readers, My Companions

To all the readers who embark on this intellectual journey alongside me, your curiosity and unquenchable thirst for knowledge inspire me to continually push the boundaries of understanding in the realm of cloud computing.

With profound appreciation and sincere gratitude,

Nikhilesh Mishra

September 08, 2023

About the Author

Nikhilesh Mishra is an extraordinary visionary, propelled by an insatiable curiosity and an unyielding passion for innovation. With a relentless commitment to exploring the boundaries of knowledge and technology, Nikhilesh has embarked on an exceptional journey to unravel the intricate complexities of our world.

Hailing from the vibrant and diverse landscape of India, Nikhilesh's pursuit of knowledge has driven him to plunge deep into the world of discovery and understanding from a remarkably young age. His unwavering determination and quest for innovation have not only cemented his position as a thought leader but have also earned him global recognition in the ever-evolving realm of technology and human understanding.

Over the years, Nikhilesh has not only mastered the art of translating complex concepts into accessible insights but has also crafted a unique talent for inspiring others to explore the limitless possibilities of human potential.

Nikhilesh's journey transcends the mere boundaries of expertise; it is a transformative odyssey that challenges conventional wisdom and redefines the essence of exploration. His commitment to pushing the boundaries and reimagining the norm serves as a luminous beacon of inspiration to all those who aspire to make a profound impact in the world of knowledge.

As you navigate the intricate corridors of human understanding and innovation, you will not only gain insight into Nikhilesh's expertise but also experience his unwavering dedication to empowering readers like you. Prepare to be enthralled as he seamlessly melds intricate insights with real-world applications, igniting the flames of curiosity and innovation within each reader.

Nikhilesh Mishra's work extends beyond the realm of authorship; it is a reflection of his steadfast commitment to shaping the future of knowledge and exploration. It is an embodiment of his boundless dedication to disseminating wisdom for the betterment of individuals worldwide.

Prepare to be inspired, enlightened, and empowered as you embark on this transformative journey alongside Nikhilesh Mishra. Your understanding of the world will be forever enriched, and your passion for exploration and innovation will reach new heights under his expert guidance.

Sincerely, **A Fellow Explorer**

Notes

Notes

Notes

Notes

Notes

Notes